Super Structures

The Science of Bridges, Buildings, Dams, and Other Feats of Engineering

MARK DENNY

The Johns Hopkins University Press
BALTIMORE

Printed in the United States of America on acid-free paper
9 8 7 6 5 4 3 2 1

The Johns Hopkins University Press
2715 North Charles Street
Baltimore, Maryland 21218-4363
www.press.jhu.edu

Library of Congress Cataloging-in-Publication Data

Denny, Mark, 1953–
Super structures : the science of bridges, buildings, dams, and other feats of engineering / Mark Denny.
p. cm.
Includes bibliographical references and index.
ISBN-13: 978-0-8018-9436-7 (hardcover : alk. paper)
ISBN-10: 0-8018-9436-0 (hardcover : alk. paper)
ISBN-13: 978-0-8018-9437-4 (pbk. : alk. paper)
ISBN-10: 0-8018-9437-9 (pbk. : alk. paper)
1. Structural engineering—Popular works. I. Title.
TA633.D425 2010
624.1′7—dc22 2009024469

A catalog record for this book is available from the British Library.

The Johns Hopkins University Press uses environmentally friendly book materials, including recycled text paper that is composed of at least 30 percent post-consumer waste, whenever possible. All of our book papers are acid-free, and our jackets and covers are printed on paper with recycled content.

Contents

Acknowledgments

The many photographs in this book, of all kinds of structures, are mostly from a handful of photographers who have traveled around the world taking pictures of buildings, bridges, and dams that caught their fancy. These skillful photographers have kindly permitted me to reproduce their work in this book. I am particularly grateful to Marvin Berryman, Don Cooper, Kees van Hest, Peter Lee, Alan McFadzean, John R. Plate, Waldemar J. Poerner, Glenn Sanchez, and Antti Olavi Sarkilahti. Thanks are also due to Kathy Castrovinci, Michael Karweit, Annemarie Spadafora, Keith Walker, and Bill Webb. The House of Commons Information Office (U.K.), the U.S. National Archives and Records Administration, and the U.S. Navy graciously supplied additional photographs. I also must thank Brent Blanchard for photographs of, and advice about, controlled demolitions.

At the Johns Hopkins University Press I am grateful (as always) to long-suffering editor Trevor Lipscombe, to copyeditor Carolyn Moser, and to art director Martha Sewall.

Super Structures

INTRODUCTION

Heavyweight Engineering

I hope that the title, *Super Structures*, gets across the subject that we will tackle in this book. You, a potential reader, naturally consider the title an important clue when deciding if a book is one that will interest you. My problem, as author, is to steer a narrow course between invoking in your mind the label "textbook" or the label "picture book." Had I called this book "An Introduction to Structural Engineering," I would have risked the former—and you might have dropped the book like a hot potato, with a slight shiver and thoughts of student days toiling over unloved mathematics in the small hours prior to an exam. People admire big structures and look upon them with a sense of wonder ("Why does this huge dam have *that* shape?"), but they don't necessarily want to have to wade through pages of math in order to get a satisfactory answer. A title such as "The Beauty of Engineering Design" might have flagged up "picture book" in your mind, giving the impression of a glossy coffee-table portfolio, long on artistic pictures but woefully short on explanations.

This book aims to explain, with technical accuracy but minimal math, why our large engineering structures—bridges, buildings, dams, and towers—are built the way that they are.[1] We have all seen amazing images of structures being demolished, of large buildings being brought down by

1. Not that I have anything against math. On the contrary, I have earned a living from mathematics and find it to be as beautiful as many of its engineering applications. But I recognize that most people do not share this view; they dislike math, or think that they are no good at it, or see it as a necessary evil. In short, it puts many people off buying books. So I include in this book the results of math analysis, but exclude the mathematical steps. Those of you who share my interest in math will be able to read between the lines and reconstruct my calculations; to this end I have included a set of 13 mathematical notes gathered together in the appendix. In addition, there are a number of technical articles and texts in the sources listed at the end of the book.

exquisitely choreographed explosions, of tall chimneys falling ("Why do they break in the middle like that?") and been impressed by the sheer scale of the structures—which becomes very clear when they fall—as well as by the skill of the demolition experts. Also it is sad, sometimes, to see a magnificent structure brought low. Equally, we have all seen how some of our large-scale engineering creations survive against the odds: slender and graceful bridges that carry loads which defy common sense, over spans that are truly astonishing ("Why doesn't that bridge collapse when the train reaches the middle?"). Those grainy World War II movie documentaries of bombed-out cities often dwell upon structures that survive unexpectedly, against the odds, such as domed churches and free-standing arches ("Why are domes and arches so stable?").

If you have ever asked yourself any of the questions that I have asked parenthetically over the last couple of paragraphs, then this book is for you.[2] If you have ever marveled at a large man-made structure—Hoover Dam, the Forth Road Bridge (or the Rail Bridge, right next to it), the Eiffel Tower, the Gateway Arch in St. Louis, etc., etc., etc.—and asked yourself questions about it, you will find many of the answers to your questions in the pages to follow. I want to share with you what I know about engineers' design skills and the expanding corpus of knowledge they have built up over many centuries. The ancient Greeks built impressive structures, but they didn't know much about arches and domes. Their successors, the Romans and Byzantines, did, but only empirically. In the seventeenth century, when Robert Hooke and Christopher Wren were thinking about rebuilding St. Paul's Cathedral in London, they had a pretty fair idea about the mechanics that stabilized arches and domes.

As the understanding of structural physics grew apace with improved building materials, the span of arches and domes increased dramatically. Railways spurred bridge development, as the iron horse sought to span continents ("How can those spindly wooden trestles carry the weight of a heavy train?"). Nowadays, nations seem to compete with each other in displaying the skills of their structural engineers—tallest building, longest-span bridge, largest enclosed volume, . . . The sheer exuberance with which we embrace large structures is a clear indication of the fascination that

2. You are not alone in being fascinated by large structures. Tourism statistics show the level of interest in these constructions: 1.2 million people visit Hoover Dam each year, while 4.1 million visit the Gateway Arch in St. Louis annually. Some other structural icons have even higher figures: 5.5 million each for the Eiffel Tower and the Taj Mahal and 40 million for the Golden Gate Bridge (including 9 million tolls).

FIGURE I.1. New materials and a good understanding of structural engineering permit buildings to take on forms never dreamed of by our ancestors. This is part of the Denver Art Museum. I thank Marvin Berryman for this image.

they hold for many people. We trust our lives with them, and yet many people have only the vaguest notions of what makes these structures work. Read on.

I begin at the metaphorical foundations and work up. So, chapter 1 is about the basic elements with which we build our large structures: stone blocks, bricks, wooden beams, metal pipes, I-beams, tubes, etc. (see fig. I.1). Different materials vary in their mechanical properties; moreover, the same material wrought into two different forms will display different strengths and weaknesses. So, a 2 × 8 plank of wood bends differently from a 4 × 4 beam. Chapter 1 presents the characteristics of building materials old and new, and their mechanical properties, in a readily digested manner (metaphorically speaking).

Chapter 2 looks at the simplest and most familiar structural component. The truss is not, in this book, a medical device for alleviating hernias but instead a structural building block that we see every time we look into a domestic roof space or at a bridge. The forces that act upon trusses have been analyzed by physicists and engineers for over a century, and an elegant theory of truss mechanics has emerged. We will dip our toes into these theoretical waters—you will learn much about how engineers calculate the forces that act upon structures—without (and I would like to emphasize this point) drowning in math. The lessons learned here will put us in fine fettle for appreciating the forces that act upon other structures, through a simple process of modeling these other structures using trusses. This approach provides a rough-and-ready method to estimate the forces that act upon all kinds of structures—from arches to dams—where a more detailed and accurate calculation would be hideously mathematical.

If you like math, you will find that my presentation gives you enough information to tackle truss theory on your own or to face a more technical introduction to the subject. The math I use is accessible to high school students—nothing more advanced than vectors and a little algebra. If, on the other hand, you are allergic to math—and many people break out in a nasty rash just hearing the word—then please don't fret, because you can read around the technical stuff without losing track of what I am trying to say. The text has been written to be read without the math (for example, all mathematical development is restricted to the appendix), so that all readers will gain insight into the forces that act upon large structures. Technical terms (italicized on first use in the text) are defined in a glossary at the end of the book.

Armed with the practical and theoretical tools of chapters 1 and 2 we look upwards, in chapter 3, to some of our tallest man-made structures: towers, steeples, skyscrapers, and other high buildings. People who build skyscrapers, for example, have solved some very interesting physics and engineering problems—problems that are unique to tall structures. I find that knowing something about the engineering design issues enhances my appreciation of these structures, and of the engineers who designed and built them.

Chapter 4 looks at the graceful arches and domes that give classical buildings, and some modern ones, their elegance and esthetic appeal. The physics of arches (and vaults; see fig. I.2) has been appreciated for centuries but is not widely known outside the engineering and physics community, so here you will see why these shapes are so much stronger than they appear to be.

I introduce simple truss bridges in chapter 2; in chapter 5 you will see much more impressive and altogether grander examples of bridges that span a lot more physics as well as a lot more space. The explosion of bridge design over the past century and a half has produced structures that are based upon many different principles; we will learn about the most

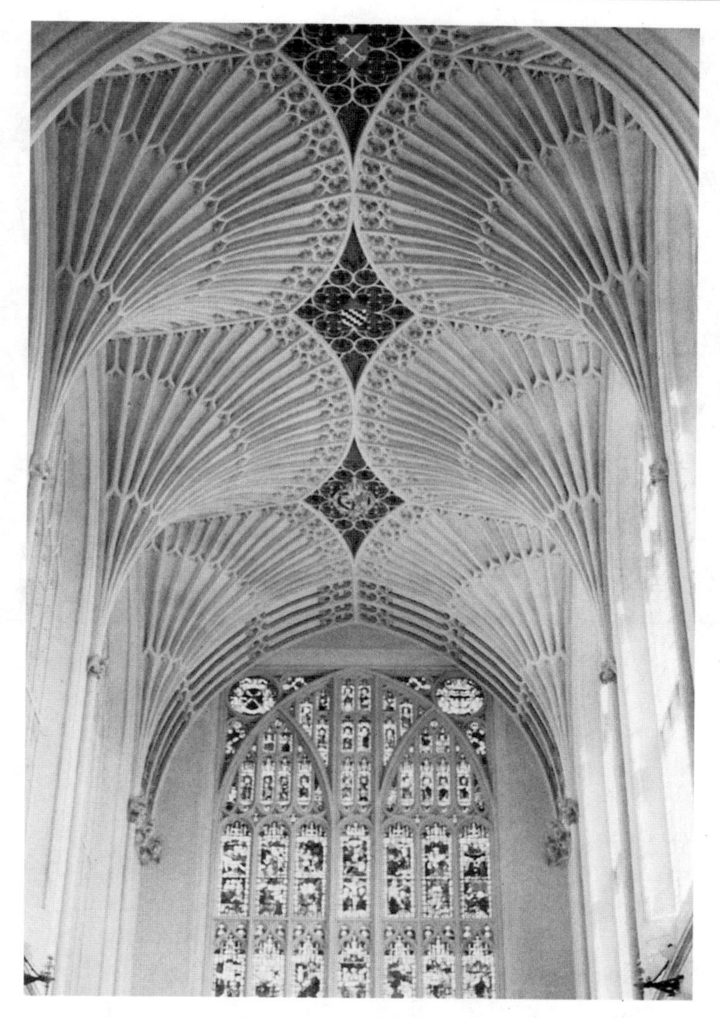

FIGURE I.2. Bath Abbey, a splendid example of English perpendicular architecture, dates from medieval times. Arches and fan vaults abound, permitting the roof to be raised, letting in light. Again, thanks to Marvin Berryman for this photo.

important of the different bridge types—illustrated with famous examples, both successful and unsuccessful—and about why there is such design diversity.

The heaviest, biggest, yet in some ways simplest of our large structures is the dam. In chapter 6 I look at the four basic dam designs and explain the physics that gave rise to them. Famous—and not so famous—dams and the

FIGURE I.3. Ceiling joists become art in this Parisian courtroom. I thank Don Cooper for this image.

stories behind them add a human dimension to these behemoths. Dams have been the pride of nations, and rightly so because they say much about our ability to understand and control the forces of nature, and about our ability to organize and carry out vast engineering enterprises.

All good things must come to an end, and the last chapter of this book deals with the death of large structures. Demolishing a tall building, particularly one that is located in a city center, is a dramatic business that is fraught with difficulty. Controlled demolition experts require an extensive knowledge of building principles, plus knowledge and experience that most building designers never need to know. We have all been thrilled by the impressive footage of large buildings and tall smokestacks that fall just where they are supposed to fall, following an exquisitely orchestrated symphony of explosions. The building drops into its own footprint (most of the time), and the smokestack falls in the direction predicted (most of the time). In chapter 7 we look behind the scenes of controlled demolitions. We also look at uncontrolled demolitions: buildings that fall down through accident, bad design, or malevolence: some fall slowly (the Leaning Tower of Pisa), but most go very quickly (the World Trade Center twin towers).

The book finishes with a short essay on our large structures as works of art. Gravity and geometry have shaped these giants, and sometimes they are beautiful. Bridges and buildings are built with esthetics in mind, as well as functionality and safety. Architecture is a mixture of art and engineering, of form and functionality, as we can appreciate from figure I.3. Much of our fascination with large structures arises from interplay between physics, geometry, and artistic appreciation. I hope that you will agree that the many photographs shown in this book—gathered from amateur and professional photographers from around the world—bring out all three facets of our long-standing love affair with our largest constructions.

CHAPTER 1

Building Blocks

It's a Material World

Stonehenge holds a fascination for many people. Its allure begins with its size and use (it probably served an astronomical function, for religious purposes) and quickly moves on to its age. We marvel at how ancient civilizations, even "primitive" societies such as the one that gave rise to Stonehenge, could construct monuments on such a scale. Delving into the details increases our appreciation: some of the heavy stones were transported by sea (probably) and by land from 150 miles distant. The lintel stones atop the pillars, shown in figure 1.1, were jointed using techniques from carpentry (mortise and tenon, tongue and groove). We can marvel at Stonehenge because it has partly survived the ravages of three and a half millennia. Whoever the people were who constructed Stonehenge, and whatever purpose motivated them to organize and execute the mammoth undertaking of building it, they surely built it to last, to send a message down the centuries for all who followed after them. And because they wanted their monument to last, they built it out of stone.

Stone and wood are two of our most ancient building materials; people used them before recorded history, and we still use them widely today. Our ancient ancestors knew as well as we do that wood is lighter to handle and easier to work than stone, and that stone lasts longer—it doesn't rot and it weathers very slowly. Everyone knows that stone and wood vary markedly in their mechanical strengths and weaknesses. This understanding, and a useful conception of structural engineering principles, was learned by humankind over many centuries. The builders of Stonehenge and of other lasting structures erected by more sophisticated ancient civilizations (fig. 1.2) lacked our knowledge of structural engineering and so were limited in what they could achieve. Both of the structures shown in figures 1.1 and 1.2 use the simplest method for spanning the gap between columns—a *lintel* or horizontal *beam*. We will see that this is not the best method and that stone is not the best material for such a construction.

Not all ancient stone constructions involve massive blocks, of course. More common, and much more versatile and easy to assemble, are structures made from smaller stones, either cut to shape or assembled from rocks shaped by nature (fig. 1.3). Today we understand much about the mechanical properties of stone, and so we know how best to use it as an engineering material. We have the technological know-how to cut and polish stone so that its beauty, as well as its strength, is best displayed. Stone is very resistant to being crushed; in modern parlance we say that it is *strong in compression*. It is relatively weak when subjected to a stretching force or pulled apart (it is *weak in tension*). Wood is quite obviously not so strong as stone. It is easier to deform—to dent, bend, and twist—and yet it has many excellent structural qualities. It is strong for its weight and is stronger in tension than compression. Because of the different qualities, we use wood as a building material differently than we use stone. This observation is evident, and would have been just as obvious to the builders of Stonehenge, though they would not have known anything about the properties of stone and wood except what they observed empirically. Our more detailed present-day knowledge comes partly from the accumulated observations and experience of many centuries, handed down the generations, but also partly from more analytical studies, both experimental and theoretical. Such analysis results from scientific, engineering, and mathematical skills that we have learned and which we now apply widely whenever we build something.

This chapter is about the materials that structural engineers utilize when doing their thing, be it constructing buildings or bridges, dams or towers. Stone and wood are two of the oldest building materials. Historically, other materials arose as we learned how to manipulate our environment: iron led to steel—an essential structural building material today. We have learned how to convert sand to glass, clay to bricks, and fiber to ropes. Within the last century or two we have added other materials: concrete, steel cables, plywood, glass-reinforced plastic, carbon composites, and so on. Let me say something about each material, and about how they differ one from another, so that we may see how they can be used as structural components.

The first iron to be employed as a structural component was *cast iron*. Nowadays cast iron is little used for structural members; it is generally considered to be not much more than an intermediate step from iron ore

FIGURE 1.1. Two views of Stonehenge, a prehistoric structure in southern England. The standing stones topped with lintel stones are 16 feet high and form part of a ring. Each stone block weighs about 4 tons; some of them were hewn from quarries 150 miles away. This phase of Stonehenge was built about 2200 BC. Images courtesy of Marvin Berryman.

to steel. In the centuries before steel, however, cast iron was utilized widely, especially during the Industrial Revolution, when iron production increased greatly. Although iron was probably not discovered in China, the process of producing cast iron was applied there more than 2,000 years before it was understood in the West. The Chinese already had developed efficient bellows, for pottery making, and so were able to generate enough heat to extract iron from ore. Cast iron is about 95% pure, most of the remaining 5% being carbon along with small amounts of other elements such as sulfur, silicon, and phosphorus. The high carbon content makes cast iron rather *brittle*, and so, as a building component, it resembles stone. It is strong in compression but weak in tension. The great advantage over stone, however, is that it can be molded into any desired shape. At the beginning of the Industrial Revolution it was utilized as a construction material in the same way as stone: the famous iron bridge in Coalbrookdale, England (fig. 1.4), was raised during the American Revolution in a small town that has become symbolic of the beginnings of industrialization. The iron foundry at Coalbrookdale is now a World Heritage Site.

Apart from bridges, cast iron was used to make cannon and shot, pots and kettles, steam engine components and wagon wheels. During the Industrial Revolution furnaces became hotter; as a result, ironmasters were able to produce purer iron with a lower carbon content, and so for the

FIGURE 1.2. Much larger and better finished than Stonehenge, this detail of stone columns and lintel forms part of the extensive system of temples at Luxor, in Egypt. Some of these constructions are as ancient as Stonehenge, while others are only half as old. To an engineer this structure is similar to Stonehenge, despite the increased sophistication. Heavy stone columns support a heavy stone lintel. Thanks to Waldemar J. Poerner for this image.

FIGURE 1.3. Two examples of old stone walls built without mortar. (a) A wall at Delphi, Greece, made by shaping stones so that they fit together closely. Image from Wikipedia. (b) A dry stone wall in Norway, made by carefully assembling natural rocks. Thanks to Alan McFadzean for this photo.

FIGURE 1.4. The iron bridge at Coalbrookdale, England. Built in the late 1770s, this bridge is made entirely of cast iron, which became widely available during the Industrial Revolution. Later iron bridges were made more economically, spanning a greater distance with less iron. Image from Wikipedia.

first time steel was manufactured in bulk.[1] Previously, steel had been a very costly, high-status material produced by hammering and working iron (in particular, to make durable swords). Steel can be worked relatively easily—it can be forged and rolled—and because it has good mechanical properties (as we will see), it grew to become a standard construction material. The key property that makes steel so much more useful to construction engineers than cast iron is that it is very strong under tension. Steel has evolved into many different forms for specialized purposes. *Steel alloys* contain varying amounts of elements other than iron (carbon, manganese, chromium, molybdenum, nickel, silicon, phosphorus, and sulfur, for example), and these impart distinct properties to the steel. Thus, stainless steel (an alloy with chromium) resists *corrosion*, while carbon steel (which also contains manganese) is very hard.

1. Symbolic of the late-nineteenth-century growth in industrialization, and of the transition from cast iron to steel, is the Eiffel Tower in Paris. This icon of France and of industry, completed in 1889, was made from 10,000 tons of cast iron and steel. I will have more to say about the Eiffel Tower in chapter 3.

Wood, stone, iron. Apart from these three, five other natural materials have been widely used as building materials: lead, rope, glass, concrete, and brick. The first three of these are of only marginal interest to us, however, and so I mention them only in passing, for completeness. Lead is a metal that resists corrosion and so has found applications in the form of roof liners and water pipes.[2] Because lead is very soft, it is of little structural use, however, and so will play no further role in this book. Another natural material that has been used for construction is fiber, in the form of rope. Rope is made by braiding certain fibers (sisal, manila, coir, cotton, and hemp, for example) and has been used to make bridges and to stiffen ships masts. During the Age of Sail, rope was produced in prodigious quantities: a large ship of the line required 30 miles of rope rigging. As a construction material, rope is useful because it is light and strong for its weight,[3] though it rots quickly. Today rope has been overtaken by steel cable, and so here it will be quickly relegated to the sidelines. When I talk about cables later in this book, I mean steel cables rather than ropes made of natural fiber.

Two more building materials that can technically claim an ancient lineage are glass and concrete. I say "technically" because, although both were known centuries ago, neither was widely used, for one reason or another. Glass was expensive to produce and was used in house construction only for small windows that formed no important structural component of the buildings they constituted—with the beautiful exception of medieval stained glass windows. Even here, though, the impressive part of the window from the structural engineer's viewpoint is not the colored glass but the fact that it was possible to build windows of such large dimensions. I will say more about this later on because constructing large windows requires a knowledge of the forces that act upon walls. I will

2. Until surprisingly recently, when the poisonous effects of lead became a much-publicized concern. As a student in Edinburgh, Scotland, in the late 1970s, I lived in an ancient tenement building that still had lead piping. The plumber who came to fix a leak hauled out yards of lead pipe and replaced it with copper pipe. (There was a government subsidy to encourage the removal of lead water pipes.) The plumber told me that to do his job in an old city like Edinburgh you needed a degree in history. He was joking, but I saw what he meant: building technology has advanced steadily over many centuries.

3. A ½-inch-diameter rope can lift a load of 230–270 pounds, depending upon which fiber and which braid is used to construct the rope. A ¾-inch-diameter rope can lift 700 pounds.

not have anything more to say about glass, however, as it is not central to our story.[4]

Concrete is very much a part of modern construction—an essential and ubiquitous structural component of almost all big engineering works. Some of you may be surprised to learn that concrete can be traced back to ancient times, but it is true. The Romans were prodigious builders and used a type of concrete. The secret of concrete making was lost with the demise of the Roman Empire and was not rediscovered until the nineteenth century. Concrete is mechanically similar to stone, but it can be more easily transported (as cement and water), molded into a desired shape, and reinforced. Concrete will, of course, figure strongly in the chapters to come.

Brick is the last of our ancient building materials. It has been around for at least two millennia and continues to be important, though its heyday was undoubtedly the Victorian era in the last half of the nineteenth century. Brick is like stone and has a number of qualities that are desirable in a building material. It is heavy and so can make solid, robust structures. Its weight also makes it a heat sink—gives it "thermal inertia" so that the temperature inside a brick structure is more even (fluctuates less) than the temperature outside—a useful property for housing. It is fire resistant and does not degrade easily (its main environmental enemy is frost). It can be—though by no means always is—aesthetically pleasing (fig. 1.5). On the downside, brick requires a solid foundation, is brittle (a definite minus in earthquake zones), and like rock, has low tensile strength. In the United States, brick making and the town of Haverstraw, New York, are inextricable. The combination of clay deposits and a ready source of water (the Hudson River) made Haverstraw the center of brick production in the late nineteenth century. At its peak the town had 41 brickworks, producing an annual total of 300 million bricks, making Haverstraw the largest brick-making community in the world.

Traditional brick making—prior to the mechanization that made Haverstraw so productive—consisted of digging clay in winter; allowing it to weather in the open air; removing any rocks and pebbles from the clay; adding anthracite (a type of coal), sand, and water; setting the mixture in

4. Glass is an important component in many modern buildings. The glass pyramid of I. M. Pei at the Louvre, in Paris, is an iconic example of glass architecture. The glass is not an integral part of the pyramid's mechanical strength—nor does the glass which clads many modern skyscrapers contribute significantly to the buildings' strength. Because glass is not a key mechanical ingredient, I do not consider it further in this book.

FIGURE 1.5. Decorative as well as functional, brick was the major building material for a century or so, until concrete came to dominate in the second half of the twentieth century. Photo by author.

molds; and kilning (heating, to dry). The Romans made flat bricks, like large tiles. Over the centuries bricks became thicker, their shape became more defined,[5] and they became decorative as well as functional. In the houses of the well-to-do in Tudor England and across Europe, exterior walls were created by inserting bricks in decorative patterns (typically *herringbone*; see below) in the gaps between timber frames (fig. 1.6). The combination of brick and timber makes for stronger walls. Bricks tend to crack, but the use of timber framing prevents such cracks from spreading catastrophically. Timber tends to warp, but the bricks help make the structure more rigid.

Bricks are cemented together (technically, *bonded*) with *mortar* in many different patterns. Some of these bonds are illustrated in figure 1.7. To

5. Some bricks are straightforward box-shaped oblongs, of standardized proportions. More typically, though, bricks have an indent on the broad face (known as a *frog*) that serves three functions. It economizes on clay, it can be decorative, and it aids the locking together of bricks and mortar.

explain the relative merits and demerits of these different bonds, I need to begin with some basic brickwork terminology. A *stretcher* is the long side of a brick, whereas a *header* is the short end. A layer of bricks is called a *course*. For a single wall (a wall that is one header thick) the most economical bond is the stretcher bond, because it wastes no bricks, except perhaps for half-bricks at the wall ends. For a double wall (one that is one stretcher thick) the simplest bond to lay is the English bond, in which alternate courses are laid with stretchers and headers visible. This method binds the double wall together in a way that is not possible if the double wall is constructed from two adjacent stretcher bond walls. If only one face of the wall is exposed (as is the case for the exterior walls of buildings), a quarter of the bricks are not visible and so can be made of inferior—less expensive—material. The English bond is the strongest double wall bond. An alternative for double walls is the Flemish bond. This is also strong and is more decorative than the English, but it is more difficult to lay. If only one face is exposed, then about a third of the bricks are not visible, so a Flemish bond wall would be

FIGURE 1.6. Half-timbered houses, with timber frames and brick or stone infill. These houses are in Hessen, Germany. Thanks to Waldemar J. Poerner for this image.

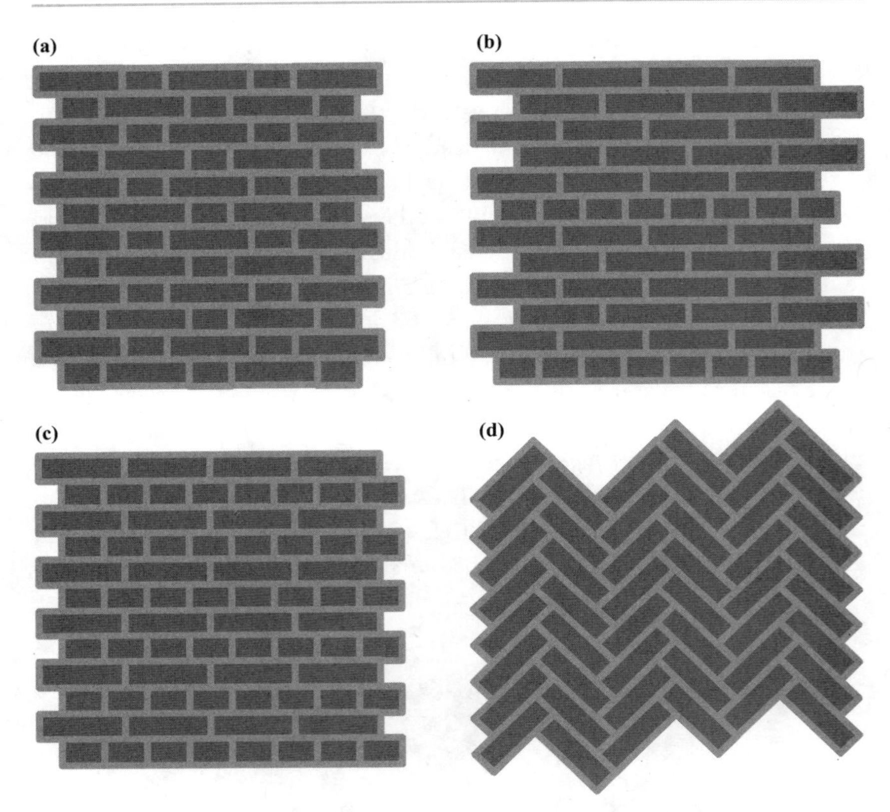

FIGURE 1.7. Some common brickwork patterns. (a) Flemish bond; (b) Scottish bond; (c) English bond; (d) herringbone bond.

slightly less expensive than an English bond wall of the same size.[6] Another common bond is the Scottish bond, which is like the English except that only one row in five, six, or seven is a header row; it is less strong than the English bond but more economical. Herringbone bond, in which the bricks are laid in diagonal rows, is used for paving, fireplaces, or other decorative features but not for load-bearing walls.

Over the last 100 years new building materials have been developed. Chief among these is the family of composites. *Composite* materials are hybrids—familiar materials that are put together in different ways to form beams, planks, and boards that are stronger, or lighter, or stiffer, than the original materials. Technically speaking, we have already encountered a

6. Less expensive in bricks, though perhaps more expensive in labor.

natural composite material: wood is made up of different types of cells that have different mechanical properties. Planks are cut along the grain because, as we all know, wood is much stronger along the grain than across it; a plank formed from wood cut across the grain would snap in two instead of bend elastically when subjected to a bending force. One of the earliest and most successful composites is plywood, which was developed specifically to overcome this weakness of wood. Thin-cut sheets of wood are cemented together with contact glue such that the grain of each sheet is at right angles to the grain of the sheets next to it. This arrangement makes the resulting plywood board stiff and strong in both directions of bending. Usually a plywood board consists of an odd number of sheets so that the grains on both sides of the board are aligned. Although plywood had been known around the world for centuries, it was only sporadically produced for limited applications until the twentieth century. It began to be manufactured for broader application when a "three-ply veneer" was exhibited at the Portland World Fair in the United States in 1905. The popularity of plywood increased markedly in 1934, when a waterproof adhesive became available, and took off[7] during World War II, when it was declared to be an essential war material.

Apart from its obvious advantage over ordinary wood—eliminating weakness across the grain—this first example of what we nowadays call "engineered wood" exhibits concomitant benefits: increased resistance to cracking, shrinking, and warping. Moreover, when the outside sheets are thin laminas of high-quality wood finishing, we obtain an inexpensive product with expensive-looking finish. Such veneers also save scarce hardwood trees or, more accurately, make the products of such trees go much further. The primary use of plywood today, however, is hidden from view. In building construction we use acres of plywood for roofing and flooring—it is difficult to imagine how the housing boom of the last century or so could have been sustained without plywood.

Another widespread and very useful composite material is *reinforced concrete* (and its close relative, *prestressed concrete*). Because concrete is weak in tension whereas steel is strong, a judicious combination of the two can yield a composite structural material that combines the strengths of both. I will have more to say about reinforced (and prestressed) concrete beams in the next section.

7. Took off literally, in one celebrated case. The successful British mosquito fighter-bomber was constructed largely from a high-grade aviation plywood.

Other new materials are the result of test-tube research rather than of re-engineering older materials: glass-reinforced plastic (fiberglass), carbon-fiber reinforced plastic, etc.[8] These new materials often appear in composite building materials for specialized purposes. For example, a composite honeycomb can be built up from a lightweight fiberglass or metal honeycomb structure sandwiched between two sheets of high-tensile-strength carbon-fiber-reinforced plastic. The result is a very light, very stiff sheet, as we will soon see.

Traditional construction materials such as wood and stone are utilized because of their mechanical properties. The strength of a stone block depends upon the type of stone. However, it also depends upon the *shape* of the block. The composite honeycomb just discussed also obtains its mechanical properties from both the nature of the constituent materials and the shape of the honeycomb. This interaction of matter and geometry makes the description of materials' strengths rather complicated. It is now time for me to describe the different forces that act upon our various construction materials, and to outline how we measure the strengths of these materials.

A Stampede of Forces

The myriad of materials just described react in different ways to the many forces they are subjected to, when *in situ* as structural components. I will get to the forces soon enough; let me here outline the ways in which engineers have learned to *describe* these forces and the manner in which they bend, stretch, and twist structural components.

We have already encountered *tensile strength*: this is the resistance of a material when pulled or stretched. More specifically, the tensile strength tells us about the maximum stretching force that a material can be subjected to without breaking. This maximum force will depend upon the thickness of material being stretched, of course, and so the force is expressed per unit area: a cable with a diameter of 1 inch will have four times the tensile strength of a 0.5-inch-diameter cable made out of the same material. So, tensile strength is usually quoted as pounds per square inch (psi) or, in the metric system, as megapascals (MPa).[9] So tensile strength (and

8. Arguably, plywood might be considered a test-tube product because a key ingredient is the adhesive that bonds the wood layers together.

9. Handy conversion factor: 1 MPa equals 145 psi.

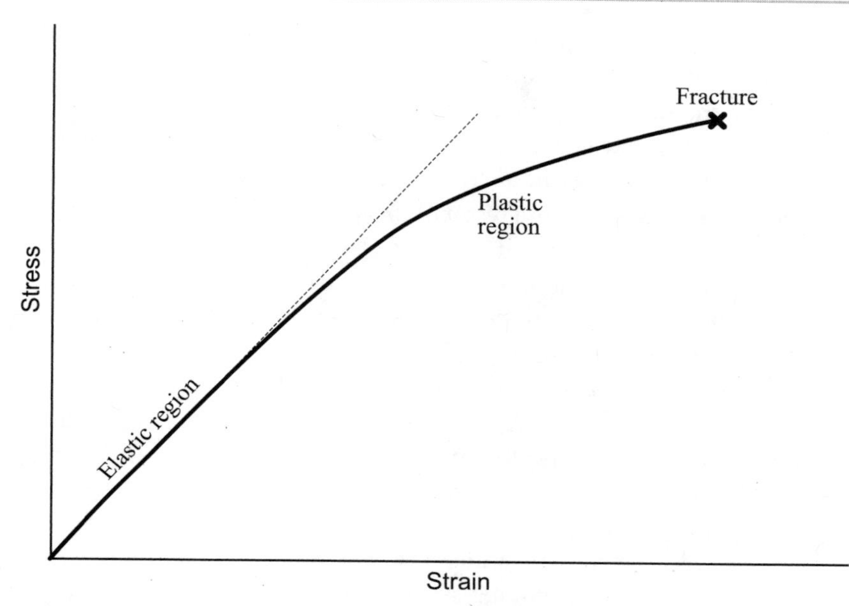

FIGURE 1.8. Typical stress-strain curve. Under small stresses a cable, beam, or column will deform elastically. For larger stresses the deformation is plastic. The stress level at which a beam will fracture depends upon the material from which it is made. The shape and length of this stress-strain curve varies widely from material to material.

the other material strengths that will be of interest to us) are expressed quantitatively in units of pressure: force per unit of cross-sectional area.

In fact, engineers quote two tensile strengths for many materials. The breaking strength is known technically as *ultimate strength*. Another, lesser tensile strength is called the *yield strength*: this is the maximum force that can be used to stretch a material without permanently deforming it. For example, if you stretch an elastic band a little and then relax the force, the elastic band returns to its original length, but if you stretch with a large force, the band will break. Somewhere in between you can stretch the elastic band enough to permanently increase its length but not enough to snap it. A similar phenomenon occurs with steel cables and with many other materials. (As you might expect, the numbers—the yield and ultimate strengths—vary greatly from one material to another.) In fact, so common are the phenomena of *elasticity* (stretching under a force and then returning to normal once the force is removed) and *plasticity* (stretching and being permanently deformed, under a larger force) that engineers observe a curve similar to that shown in figure 1.8 for many different materials. As stretching force is increased—say a nylon rope is being pulled—the material first stretches elastically, then plastically as force increases, and then it snaps.

These curves are known as *stress-strain* curves. "Stress" means the applied force (per unit area) and "strain" denotes the amount of deformation—stretching—that occurs as a result of the force. Once again, elastic deformation is temporary: a material stretched elastically will resume its former length when the stress is removed. Plastic deformation is permanent: removing the stress leaves the material altered. In figure 1.8 you can see how our nylon cable acts under tensile stresses. A similar curve applies for steel cable, and for bricks, stone blocks, concrete beams, and other materials. The extent of elastic/plastic regions varies hugely from one material to another, and likewise the ultimate strengths of materials diverge widely. So, the plastic region for nylon is large, but it is small for cast iron. The fracture stress for steel cable is way, way larger than that for rubber.

We have also already encountered *compression strength*, which is the resistance of a material to compressive force. Compressive strength is the squashing force per unit area that a material can resist without crushing. As with tensile strength, we have to be careful about what we mean: a material that is subjected to compressive stress (say, by putting it in a vise) may deform in different ways. If it is thin—for instance, a knitting needle that is compressed along its length—then it may buckle and return to its original shape when the compressive force is removed. Or it may stay permanently bent. If a thicker material—perhaps a block of shortbread,[10] or steel, or soap—is squashed, then it may deform by crumbling or cracking (fracturing) or spreading out sideways (plastic deformation). In brief, materials respond to crushing stress in as many and varied ways as they respond to tensile stress, and engineers have to find ways to describe these responses. The compressive strength of shortbread is not of interest to structural engineers, but the compressive strengths of steel, concrete, wood, and brick are of great interest. They must be known accurately so that buildings, dams, and bridges can be safely constructed. So, in the technical literature you can find innumerable tables of numbers representing compressive (and tensile) strengths of different materials. Structural engineers

10. I am not suggesting that we use shortbread as a building material—there are several objections that will immediately occur to you—but shortbread *is* a material, and it does have a (small) compressive and tensile strength.

can calculate from these strengths how high they can make a building, or how long a span they can build a suspension bridge, and so on. We have come a long way since the "five-minute rule" of the ancients.[11]

There are many other types of forces that can act upon a structural material. *Bending strength* (also known as *flexural strength*) describes how much a material can be bent before it fails. *Shear strength* describes the resistance of a material to shear forces (explained shortly). *Torsional strength* tells us how well a material resists being twisted. As before, we must be careful how we quantify these different forces and the different ways that they deform materials. Here is a list of material properties, some of which we have discussed and some of which are obvious:

elasticity	hardness	stability
plasticity	toughness	creep
malleability	brittleness	fatigue
ductility	stiffness	corrosion

Ductility is the ability of a material to stretch or bend before fracturing, when subjected to a tensile force (it is the opposite of brittleness); *malleability* is the equivalent property for materials subjected to a compressive force. Lead and other metals are both ductile and malleable, glass (except near its melting point) is neither. *Creep* refers to the gradual deformation of a material when force is applied. I do not propose to expand further upon these properties of materials. The main thing that I want you to take away from this section is the variability of materials and of the effects that different forces have upon them.

These effects depend upon material shape as well as composition. Structural engineers often describe a given material not just in terms of, say, tensile strength but also in terms of various other parameters. Thus, there are three parameters that engineers use to describe the elastic properties of a material. *Young's modulus* (usually denoted E) describes tensile elasticity; a material that will not easily stretch has a large Young's modulus.[12] *Shear*

11. The five-minute rule: if a structure doesn't fall down after five minutes, then the design is good.

12. Young's modulus is the ratio of stress to strain; in other words, it is the slope of the elastic part of the stress-strain curve shown in fig. 1.8. The fact that this part of the curve is straight reflects the well-known *Hooke's law* of deformation.

TABLE 1.1. Density and crushing strengths of various construction materials

Material	Density (lbs / ft³)	Crushing stress (psi)	Max. height[a] (ft)
Adobe	100	500	720
Timber	60	2,000	4,800
Brick	120	4,000	4,800
Concrete	150	5,000	4,800
Granite	160	15,000	13,500
Steel	490	50,000	14,700

[a]The maximum height that materials can be stacked (say, forming a wall) depends upon density as well as crushing strength.

modulus describes the rigidity of a material,[13] and *bulk modulus* tells us of a material's volumetric elasticity (a three-dimensional version of E). I mention these moduli only to reinforce the notion that there are many ways to deform a material; consequently, an engineer must know many parameters for a material when he or she is considering its mechanical properties.

One last example of deformation variability: some deformations are transient or cyclic—they vary with time—whereas others are permanent. An automobile tire deforms cyclically every time it rotates. The specially formulated composite material that constitutes the tire is designed to withstand many such cyclic elastic deformations (perhaps 100 million during the tire's lifetime). A floorboard undergoes elastic deformation each time you walk over it (unless you fall through the floor), and it must be able to withstand this deformation many times. Indeed, the floorboard is made of wood precisely because we know that wood is capable of withstanding such transient forces without permanent deformation. Bricks that form a wall, on the other hand, are subjected to a permanent compressive load (the weight of bricks above), and so they must be able to withstand strong compressive forces without deforming over time. Indeed, it is precisely because bricks can do this job that they are used to make walls.

Table 1.1 lists the densities and typical crushing stresses (the compressive stresses required to cause a material to fail) for several building materials. Tensile strengths are generally smaller than compressive strengths (concrete has a tensile strength that is roughly 10% of its compressive

13. Technically, shear modulus describes the ease with which a material of a given shape can be deformed without changing volume.

strength) but not always: some types of steel have tensile strengths that exceed compressive strength. The numbers shown here do not necessarily apply to other materials that you may consider to be similar. Thus, the crushing strength of sandstone is less than 40% that of granite. From the density and crushing stress figures in this table, we can work out the maximum theoretical height of a building made from each material: these heights are shown in the last column. I say "theoretical" because for real buildings we assume generous safety factors that allow a wide margin for construction errors and material variabilities, so real buildings will never be built as high as the numbers of table 1.1 suggest. But these numbers do tell us something about the suitability of different materials for making big structures that are subjected to large compressive forces. Construction engineers will make similar (though much more detailed) calculations concerning tensile, shear, bending strengths, etc., before deciding upon which material to use for which component of a large structure.

Beam Me Up

In the previous section we saw how different structural materials have different strengths and weaknesses. The forces that act upon structures are varied, and their description is complicated. One of the complications is that the strength of a given mass of material, say concrete, depends upon its shape. In this section I want to give you a feeling for this shape dependence, which naturally leads me into the complicated topic of beam theory. However, I have a problem. Beam theory, while it is well understood mathematically, really is complicated, and my purpose in writing this book is not to write a text on engineering mathematics.[14] Most of you, I suspect, would fall asleep reading such a text, and I, frankly, would rather undergo

14. Most introductory engineering mathematics texts will provide you with a firm grounding in so-called "simple" beam theory. You can believe the mathematics —this theory has been developed and tested over three centuries and provides a reasonable quantitative description of the stresses that develop within a beam when it is loaded—but don't believe the name: there is nothing simple about it. And even this theory is only an approximation. The twentieth century produced a "generalized beam theory" that is a better approximation to the true nature of beams. The generalized theory does away with one of the assumptions of the simple theory and accounts for the cross-sectional distortion (*transverse shear deformation*, in engineer-speak) that is known to occur when a beam is bent or sheared under a load. Beam theory remains a topic of active research today.

root canal work than write it. Here, I will instead give you a flavor of beam theory; a few diagrams and a descriptive text will get across the essential points that you will need for tackling the rest of this book. In the next chapter we will get into more detail regarding the theory of trusses because truss theory is mathematically much simpler, and it carries us a long way; we will even be able to gain a quantitative appreciation of beam theory by considering beams as trusses. But I am getting ahead of myself: first, we need to understand the relationship between shape and strength.

Consider figure 1.9a. Here we have a horizontal beam (of oak, steel, or concrete) that is in the shape of a solid oblong. The beam is loaded—subject to forces—as shown. A large force bears down on the center of the beam from above, while two smaller forces, acting at the ends, hold the beam in position. Perhaps the beam is the lintel above a large doorway, in which case the end forces represent the doorposts that hold up the beam, and the large downward force is the weight of wall above the door. It is not difficult to see from this diagram how the forces are going to deform the beam: it will bend down in the middle. This bowing deflection will be small if the beam is strong enough to do its job, but it is present. The uppermost surface of the beam will be slightly compressed as a result of the bending, while the lower surface will be slightly stretched. In other words, the upper part of the beam is in compression while the lower part is in tension.

This simple observation takes us forward quite a ways because as soon as we realize that the forces on two sides of a deformed beam can be very different, we see that there are a number of practical consequences. A sturdy beam will have to be strong in both compression and tension. We saw in the last section that concrete is weak in tension, though it is strong under compressive forces. This weakness in tension limits the length of a concrete beam. It is obvious that the beam in figure 1.9a will bend more if it is thinner or longer. A thin or long concrete beam will break under a load, even under its own weight, at the bottom center.[15] An oak beam will probably fail at the top center because wood is stronger under tension than under compression. Another consequence of our observation: if the top

15. Bricklayers take advantage of this weakness under tension when they want to cleave a brick in half. The brick is put up on two supports at the ends, and a stone chisel is placed half way along the upper surface, and struck sharply with a hammer. The brick breaks neatly down the middle not because the chisel forced its way through the upper surface, but because the brick fractured at the lower surface, halfway along, because brick is weak under tension.

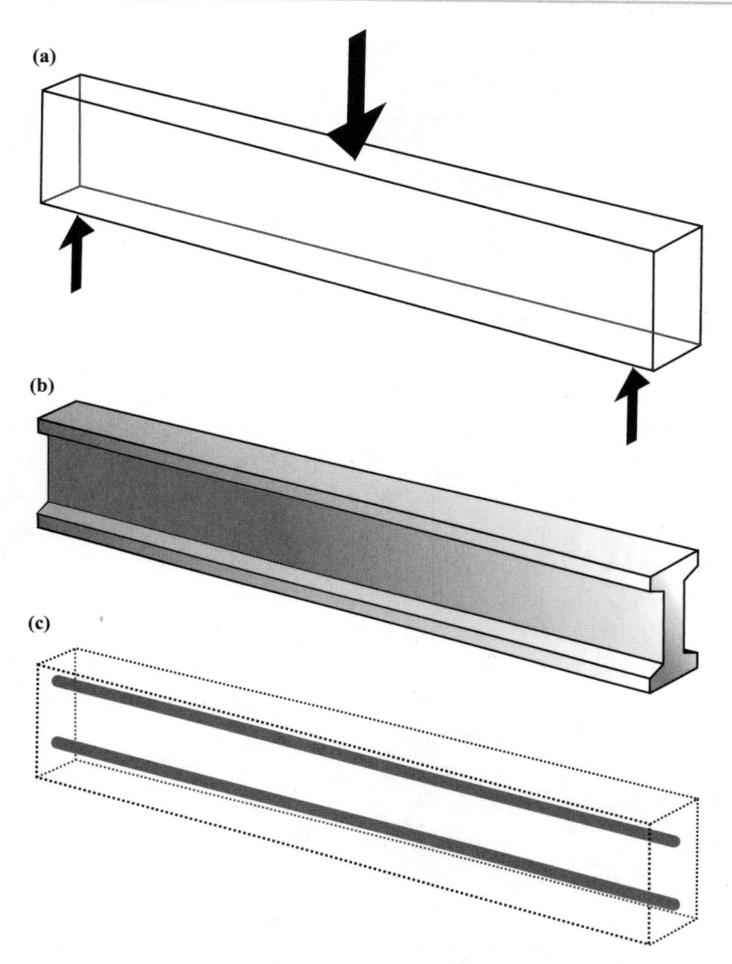

FIGURE 1.9. Bending force applied to a beam. (a) A solid beam is subjected to forces that will cause it to bend. (b) An I-beam of the same external dimensions will resist bending almost as well and will be significantly less heavy. (c) A concrete beam that is reinforced, and perhaps prestressed, with metal bars is strong under both compression and tension.

surface of the beam is in compression and the bottom surface is in tension, then there must be a layer near (or at) the center of the beam that is subjected to neither tensile nor compressive forces. This layer is termed the *neutral layer* by structural engineers. Because the center feels little force, it does not contribute much to the strength of the beam (assuming that the beam is expected to be subject only to bending forces); to save weight and expense, therefore, we may as well remove the center, leaving us with the I-beam of figure 1.9b. This beam (let us say it is steel) has the same size of upper and lower surfaces as the original beam and so is almost as strong under a bending load. Yet it is significantly lighter and thus less expensive. I-beams are ubiquitous in buildings because of their high (bending) strength-to-weight ratio.[16]

I said earlier that concrete was weak in tension and also that reinforced concrete addressed this problem. You can see in figure 1.9c how reinforced concrete beams are constructed. Steel rods (reinforcing bars, or *rebar*) are placed in a form which is then filled with liquid concrete. When the concrete sets, the bar ends are cut off leaving the composite structure shown. The rebar has a roughened surface so that it is gripped by the concrete and does not slip when the beam is loaded. You can see that the beam of figure 1.9c, when subjected to the bending load of figure 1.9a, will do better than a pure concrete beam. The concrete will deal well with the compressive force at or near the upper surface because concrete is strong—deforms very little—under compression. The lower steel rebar will provide tensile strength for the lower part of the beam. So, when you see a concrete beam spanning a large space (being used, for example, as the beam above a double garage), then you know for sure that this beam is reinforced. An unreinforced concrete beam would break.

What about prestressed concrete? This is the same as reinforced concrete except that the rebar is stretched mechanically when the concrete is being poured, and this external tension is not released until the concrete has set. The result is that the rebar inside the beam is under tension, pulling the ends of the concrete beam together, so that the concrete in the beam is under a strong compressive force even before it is subjected to external loading. Now if we subject the prestressed beam to the bending force of figure 1.9a, the lower surface may not even be subject to a tensile force pulling the concrete apart because the tension due to the bending load is more than countered by the compression due to prestressing. You can see that prestressing a concrete beam makes it even stronger than simply reinforcing it (assuming that the rebar is not so highly stretched that it is close to breaking).

To back up the ideas behind figure 1.9, consider figure 1.10. Imagine a

16. Another example is hollow metal tubing, which is not as strong (as resistant to bending or buckling) as a solid rod of the same diameter, but is still pretty strong and has a much higher strength-to-weight ratio.

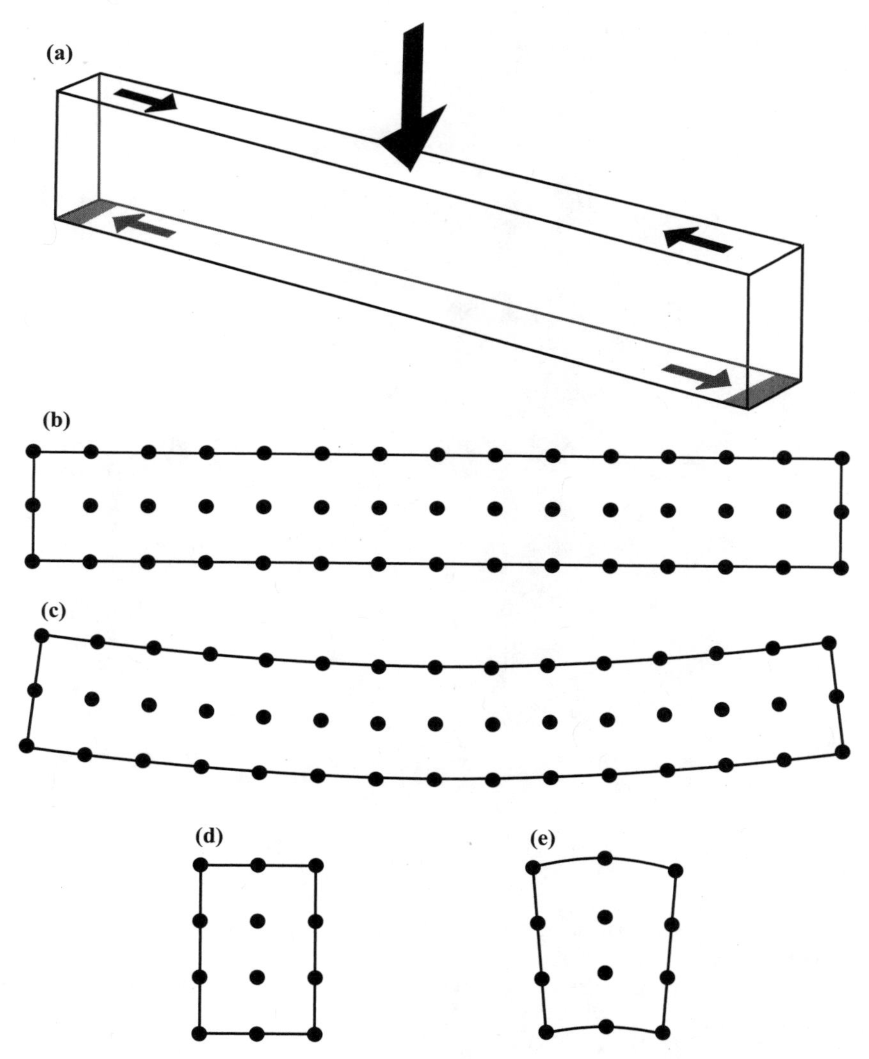

FIGURE 1.10. Beam distortion. (a) A beam supported at both ends (shaded rectangles) is subjected to a force (vertical arrow) that distorts the beam: the upper surface is compressed and the lower surface is stretched, as indicated by the horizontal arrows. (b) The unbent beam, viewed from the side. (c) The bending beam, viewed from the side; note the compression of the upper surface and stretching of the lower surface. The "neutral" layer is neither compressed nor stretched. (d) The undistorted beam end. (e) The beam end as it appears when the beam is bending.

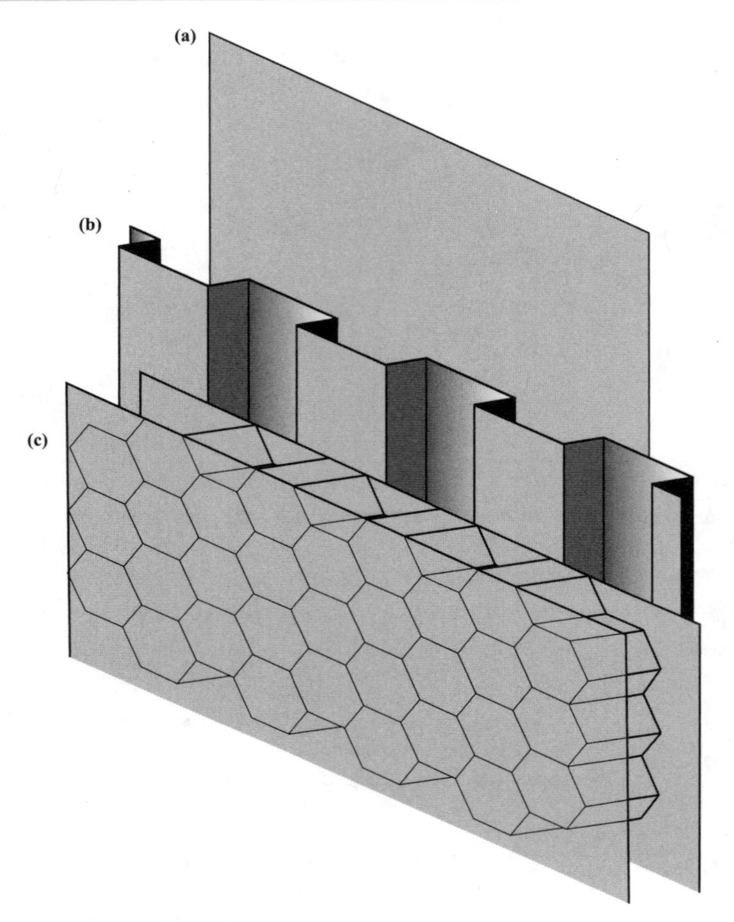

FIGURE 1.11. Strength depends upon shape. A plane sheet (a) can be easily bent in two directions. A corrugated sheet of the same thickness (b) can be easily bent in only one direction. The honeycomb sandwiched between two sheets (c) cannot be easily bent at all.

grid of dots painted all over the surface of the beam. This grid becomes distorted as the beam is bent, as shown, and these distortions indicate how the different parts of the beam are stretched or compressed under a bending load. Also note from figure 1.10e how the beam cross section becomes distorted. Think of an elastic band being stretched: it becomes thinner. Same here: the lower end of the beam (under tension) becomes thinner while the upper end fattens. This distortion greatly complicates the

TABLE 1.2. Stiffness and strength of a honeycomb composite compared with a plane sheet of metal

Material	Thickness	Weight	Stiffness	Strength
Plane sheet	*t*	1.00	1	1.0
Honeycomb composite A	2*t*	1.04	7	3.5
Honeycomb composite B	4*t*	1.06	37	9.2

detailed mathematical theory of beams, and so, having pointed it out to you, I now drop it like a hot potato.

The same idea that led to I-beams also leads us to corrugated sheets. In figure 1.11 we see a plane sheet (of steel or aluminum) and a corrugated sheet. The corrugated sheet is stiffer than the plane sheet if we try to bend it about an axis perpendicular to the corrugation. A little thought will convince you that when this corrugated sheet is subjected to a bending force, different parts of it are in tension and compression. The extra thickness of the corrugated sheet (due to the corrugations) makes it stiffer and better able to resist the bending force. Replace the corrugated sheet with a solid sheet that is as thick as the corrugation and you will undoubtedly obtain increased resistance to bending but at the cost of vastly increased weight (and price). As with the I-beam, the corrugated sheet has a much higher strength-to-weight ratio than the equivalent solid material.

In figure 1.11c we see a modern version of a corrugated sheet. Honeycomb composites, discussed in the last section, are very light and stiff. The advantages of a honeycomb "sandwich" over a plane sheet of metal can be seen in Table 1.2. These numbers, taken from the website of a honeycomb composite manufacturer, show (in appropriate units) impressive gains in strength and stiffness for little gain in weight. In part, these improvements are due to increased thickness and, in part, to the honeycomb structure. One advantage of the honeycomb sandwich, compared with the corrugated sheet, is that the sandwich is stiff in all directions. A thick honeycomb does not even have to be inside a sandwich to be strong; an example is the steel honeycomb road surface of figure 1.12.

Corrugation is less expensive, however, and for many applications the extra bending strength is needed in only one direction. For example, the security door shown in figure 1.13 is backed by strong posts that provide stiffness in the direction where the corrugations are weak.

Thus far, my qualitative explanation of beam theory has involved only

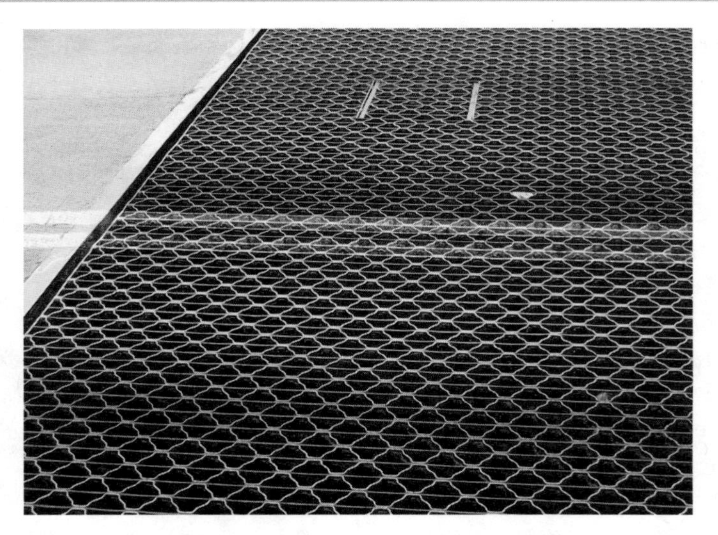

FIGURE 1.12. A steel honeycomb road surface on a swinging bridge. The honeycomb structure makes the road surface stiff, with a high strength-to-weight ratio.

bending forces. We have seen that other types of force can apply to structural materials. Some of these forces are illustrated in figure 1.14. Since we have already encountered tension and compression, I need not comment further upon figure 1.14b and c. The shear force (fig. 1.14d and e) requires a little more effort to explain. You can see that it is similar to bending in that it arises from applying an external force perpendicular to the beam length, but there the similarity ends. Bending and shear produce quite different deformations in the beam, as you can see by comparison with figure 1.10c. In fact, the distortion shown in figure 1.14d glosses over a complication: it is two-dimensional. Viewed from the beam end, we would see that a beam in shear is distorted in a quite complicated and unintuitive way. Within the framework of simple beam theory we can ignore these complications because the practical effects are usually small.

Simple Beam Theory Results

The cantilever beam is a very simple structural component. Simple to visualize, that is (fig. 1.15) and simple compared with many real-life structural components, but even this elementary case is hard to analyze mathematically. If you would like an idea of the complexities of simple beam theory, I refer you to the standard formulas and their explanation in note 1

FIGURE 1.13. A corrugated steel door is resistant to bending in one direction; in this case it cannot be bent easily about a vertical axis. Strong posts behind the door will resist bending about a horizontal axis. Result: a lightweight door that is not easily deformed by force.

of the appendix. We will not need the results of note 1 in later sections, but they serve to convey the depth of math analysis required for beam theory—and so motivate us to seek a simpler method of analysis (truss theory, in the next chapter).

Another "simple" case is that of buckling. In figure 1.16 we see what can happen when a vertical beam (a column) is put in a vise or loaded with a great weight. Thin columns may *buckle*, while thicker ones are more likely to fracture by shearing or to be crushed.[17] The force required to make a thin column buckle can be calculated from beam theory—it was one of the first examples to be evaluated mathematically, centuries ago. This force is discussed in note 2 of the appendix.

17. Beam theory calculations tell us the *critical column length*, above which a column will fail by buckling rather than by compression.

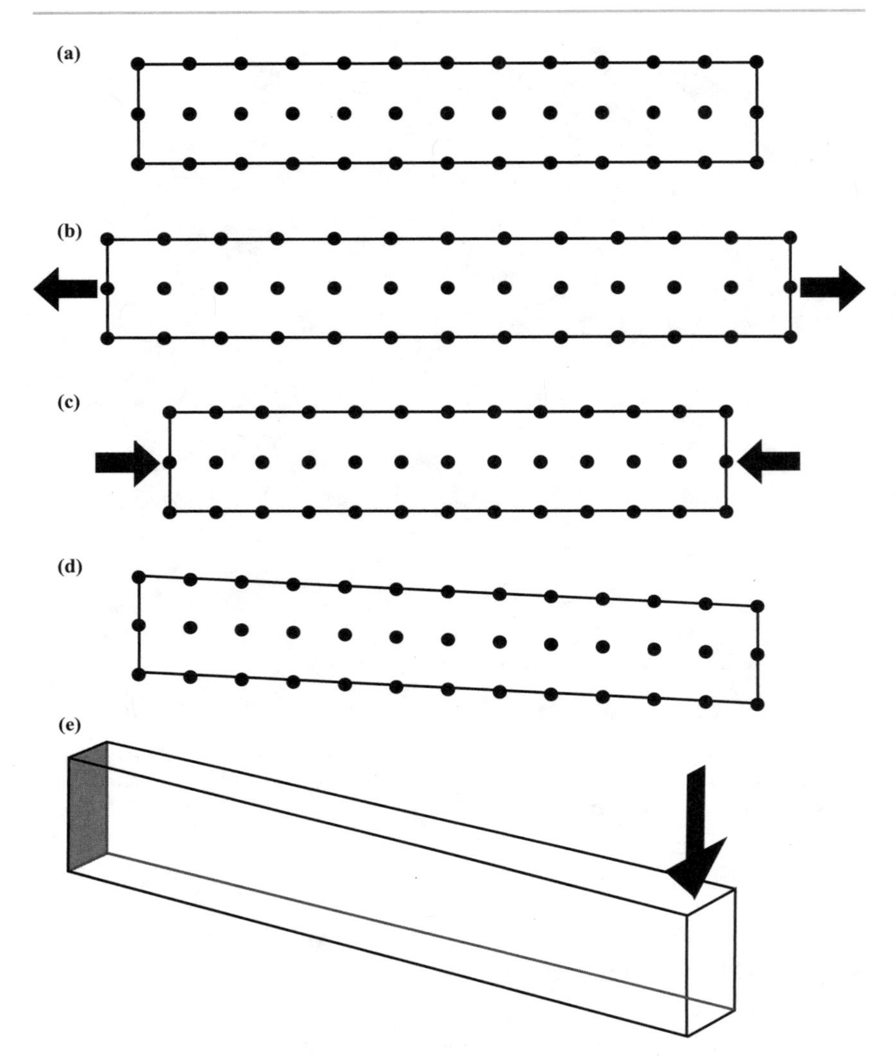

FIGURE 1.14. Some forces that apply to beams: (a) undistorted beam, viewed from the side; (b) stretched (tensioned) beam; (c) compressed beam; (d) sheared beam; (e) beam that is fixed at one end (shaded) with a force applied at the other end (vertical arrow), producing shear. For tension, compression, and shear forces, molecules of the beam that are initially in the same vertical plane stay in the same plane. This is not the case for bending forces.

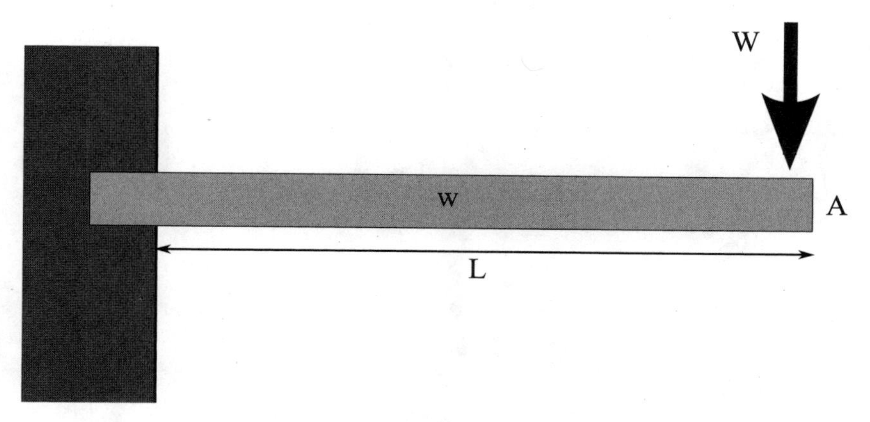

FIGURE 1.15. A beam of weight *w*, fixed at one end, is subjected to a load *W* at the other end. For specified dimensions and weight we can calculate how a beam deflects due to the load. Here, beam length is *L* and cross-sectional area is *A*.

For the stone columns used in classical buildings to support beams or arches (as in the elegant example of fig. 1.17), there is a rule of thumb that ancient builders learned to respect. For the column to safely support the load above it, the line of force must be near the center of the column base. If the line of force falls outside the base, the column will fall sideways. For safety, the line of force should lie within the middle third of the base (hence, this rule is known as the *rule of the middle third*) in both dimensions, so the line of force should lie within the middle ninth of the base area. For our column in a vise (fig. 1.16), the force was vertical, with the line of force right down the center, and so we only needed to worry about whether or not the column was strong enough to resist the force. For real stone columns the loading is such that the force is rarely vertical, and hence the need for our middle-third rule. More on lines of force later.

While arguably more beautiful, and architecturally more elegant, the column and beams of figure 1.17 are structurally very similar to the column and lintel of Stonehenge (fig. 1.1) or of the ancient Egyptian structure shown in figure 1.2. Column-and-beam structures do not take advantage of the resistance to sideways slipping of the ground upon which they stand (the ground's horizontal strength) but only upon its resistance to compression (the vertical strength). This means that forces acting upon a structure are often much greater than the load that the structure is intended to

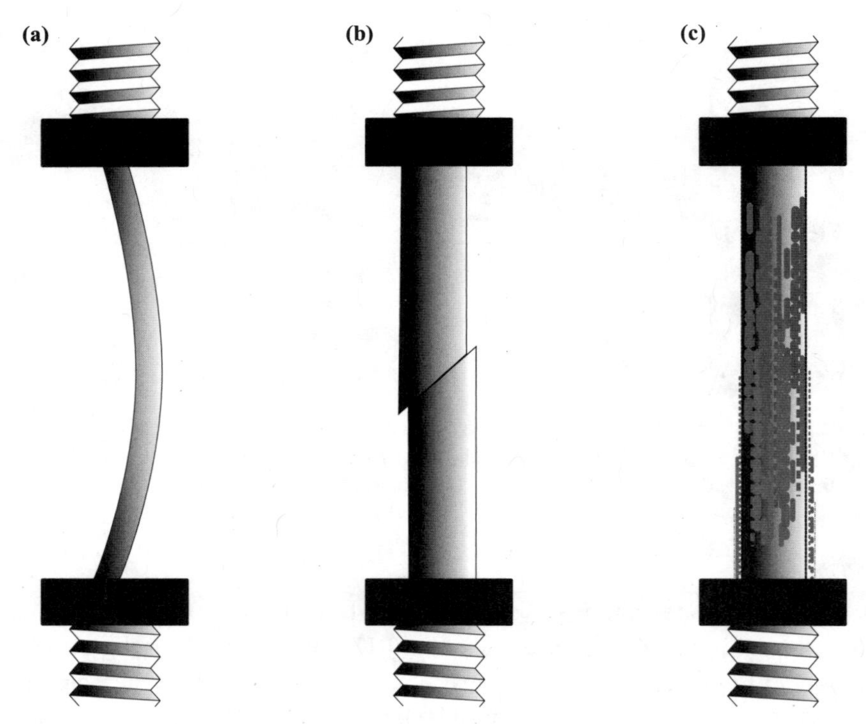

FIGURE 1.16. Under compression a thin column will buckle (a), whereas a thick column may shear (b) or be crushed (c). Which of these distortions occurs depends upon the column material.

support. As a result, the spans between columns tend to be short and the lintels or beams tend to be thick and heavy. Architecture of this sort is referred to as *trabeated* and is a sign that the builders did not understand beam theory and had not developed the more economical and stronger technique of spanning large spaces with arches. Indeed, one expert has described the ancient Greeks as "decidedly timid" in their construction of stone lintels. They erred on the side of safety, and so their structures look heavy, though beautifully executed. We will see in chapter 4 how the arch opens up space, spanning larger gaps with less stonework, and thus expanding architectural possibilities and enhancing the functionality of buildings. Building with arches requires at least an empirical knowledge of the forces that act on structural components and knowledge about the strength of building materials that are used in the construction.

FIGURE 1.17. This attractive row of columns, in Guimaraes Castle in Portugal, supports a long beam. Thanks to Waldemar J. Poerner for this image.

Putting It All Together

What of the connectors that hold together the beams, columns, rafters, trusses, etc., to form our large constructions? Nails and mortar have been used for at least 2,000 years, and screws, bolts, adhesives, and rivets for at least a couple of centuries. Nails, screws, and bolts rely upon friction to hold the show together. If you were a nefarious magician with the power to snap your fingers and abolish friction from the world, you would hear the sound of collapsing buildings within a second or two of the snap. Your own house would become part of the rubble on your street and city. Nails would fall out of rafters, drywall would fall off joists, screws would pop out, and nuts would slip from bolts. Old-fashioned metal bridges, however, might escape destruction because many of these are held together by rivets. Newer metal connections are made by welds; these also would remain intact. Plywood and laminates would not fall apart because adhesives do not rely upon friction. But most wooden constructions would disintegrate.

The world, fortunately, has not produced any such nefarious magicians so far, and friction can be relied upon to hold wood and metal together. Nails that were hammered into place in medieval Europe still hold together the beams of timber frame houses. Anyone who, like me, has installed a hardwood floor by hand-nailing (1,500 nails, as I recall) becomes acutely aware of the power of friction. The nails that I used were specialized for the purpose of connecting hardwood flooring to a wooden subfloor, and this specialization is typical of nails in the modern world. Nails were handmade until the eighteenth century, but then the Industrial Revolution gave us the ability to mass-produce identical nails by the million. Different materials require different types of nails, however, and by the end of the nineteenth century nail evolution had flowered into dozens of different species. Today there are hundreds, as a visit to your local hardware store will show. The same evolution is evident in screws. The screw is a machine well known to the ancients,[18] but the familiar connector that holds your deck together is relatively recent because screw production requires precision engineering. As with nails, screws have evolved into hundreds of specialized types.

Bolts differ from screws in that bolts are not designed to be turned. They fit into a hole that is of larger diameter and are held in place by a nut. The strength of a bolt depends upon the steel that constitutes it, of course, but also upon the thread of the bolt and nut and upon the torque that is applied to hold them together. Both screws and bolts are usually under tension when in place, though it is notoriously difficult for engineers to *measure* the tension without retorsioning. Rivets (fig. 1.18) are like unthreaded bolts. A factory-made head is at one end of the cylindrical shaft; the other end (the tail) is flattened when the rivet is in place and still red-hot (and therefore soft and workable). The resulting double-headed rivet shrinks as it cools, making a tight connection. It can hold together two metal beams or sheets that are in tension (i.e., subjected to a force pulling the beams apart), but it is better at resisting shear loads (a force pulling the beams sideways). Rivets predate welding and modern bolted joints and have successfully held together many of the large structures that were first produced as icons of the modern age, such as the Eiffel Tower in Paris and the Forth Rail Bridge just outside Edinburgh, Scotland. Rivets are still widely used in aircraft manufacturing.

18. The Archimedes screw was a device used in antiquity for raising water from a river.

FIGURE 1.18. Steel beams and trusses are held together by rivets. Photo by author.

I have already mentioned adhesives in the context of laminates such as plywood. Modern adhesives are capable of holding together any two solid objects.[19] The variety of adhesives that are widely available has flowered as profusely as the variety in nails and screws. One of the main consequences of this flowering for today's building construction concerns the improved economic use of timber. We saw something of this economic gain when considering plywood veneer, but the importance of strong wood adhesives today has spread beyond plywood. Many of the wooden beams, and even planks, that you see around you are not cut from a single piece of timber but are formed from many smaller pieces held together firmly via adhesives. My house is a typical older (for North America) wooden dwelling, with its mixture of 50-year-old roof beams each made from a single piece of fir and more recent beams made from smaller planks glued together. This use of adhesives permits more useful timber to be extracted from each tree that is felled, because shorter or smaller pieces that would in earlier decades be discarded are today glued together to form larger planks or beams. So good are the modern adhesives, indeed, that a carpenter need pay little

19. Not quite true, of course, and you will no doubt have your own favorite exception. Personally, I am unaware of any adhesive that will successfully bond ice to Teflon, but then I can think of no very good reason why I would want to make such a bond.

attention to the fact that his wood is composite: a given length of a composite 2 × 2 is as strong as the same length of a 2 × 2 formed from a single piece of timber.

Mortar is the glue that holds together stone, brick, or concrete blocks. (Indeed, modern mortar *is* concrete, pretty much, if you add gravel.) Lime mortar was used for at least 6,000 years; it was known to the ancient Egyptians. It was made by burning limestone in a kiln to yield quicklime, which was then slaked (mixed with water) and mixed with sand. It bonded quite well, once set, but it set extremely slowly. A Roman variant, *pozzolana,* a sandy volcanic ash that was mixed with lime, set quickly and was more like modern concrete; like concrete it even set underwater. Modern Portland cement mortar dates from the first half of the nineteenth century.[20] It is made by heating limestone with certain secondary ingredients (such as clay, shale, sand, and iron ore) and grinding the mixture into a powder. We now understand the chemistry of cement and concrete very well. Cement is graded into five different types, which vary in strength, resistance to corrosion, and cost. As we saw earlier, concrete has mechanical properties similar to stone, its main mechanical advantage being that it is pure and uniform, with no planes of weakness, and so its strength is predictable.

SUMMARY: In this chapter, we have learned about the strengths and weaknesses of traditional building materials such as stone and wood, and of their modern equivalents, concrete and steel. We have surveyed the forces (tension, compression, bending, etc.) that act upon these materials and the relevant material properties (elasticity, plasticity, etc.). An overview of beam theory showed us that some materials are better than others for certain techniques of building. We saw that the limited range of construction materials available many centuries ago (and, perhaps more importantly, the limited understanding of construction engineering concepts) limited the methods of construction.

20. Named for the Isle of Portland, off the south coast of England, where there is an old quarry that has produced rock for centuries. Portland cement sets into concrete that much resembled this rock.

CHAPTER 2

Truss in All Things High

Members of the Assembly

From clay we make bricks, and from bricks we make walls. Similarly, from the elementary materials described in chapter 1 we make structural trusses, and from trusses we make bridges or roofs (or gangways, cranes, fences, furniture, . . .). The main subject that I will address in this chapter is truss theory; it will provide us with a theoretical analysis tool that will serve us well for the rest of the book. We begin our quantitative analysis of big structures with truss theory, rather than with the beam theory that I skirted around in chapter 1, because truss theory is much, much easier to understand. We can use truss theory to get a feeling for the forces acting on non-truss structures, such as beams, towers, and dams. So, truss theory will set us up with a mental toolkit that we will put to use in later chapters.

The most immediately obvious, and historically important, application of structural trusses is to span spaces with either bridges or roofs. We have all seen trusses in the wooden beams of a house or church, and in the network of metal bars and rods that form many bridges. So, part of this chapter will be devoted to bridges and roofs and to showing how an understanding of truss forces permits construction engineers to design them. We will see that roof truss design evolved over the centuries, allowing builders to span larger and larger spaces, and that these evolving designs came more and more to resemble arches, structurally speaking (though their appearance could be very different). By the nineteenth and twentieth centuries, bridge engineers were called upon to span much larger distances than in earlier times, and these distances became too much for simple truss structures. In this chapter we will touch upon short-span bridges and will examine the astonishing long-span bridges of the past 150 years in chapter 5.

To a student at the beginning of a career in mechanical or civil engineering, a truss is an abstraction—a theoretical construct invented to simplify calculations. Later it becomes real. Physical structures are built from actual trusses, but first the student (and the reader) needs to understand the forces that will act upon large structures. We surveyed these forces in a qualitative way in chapter 1, but what we really need is a convenient, intuitive method of quantifying the size and direction of forces. Here, the abstract truss performs its greatest service. This book is all about cutting through confusing details (be they mathematical or engineering technicalities) to reveal underlying physical principles, and truss analysis does the same thing. So, the best way that I can begin this chapter is to get you up to speed on the theory of trusses, and how they are put together, like Meccano or Erector Set toys, to form large and stable structures.

A truss is a structure that is assembled from rigid bars, called *members*. To simplify engineering calculations about structural stability, both the structure and the members are idealized. These idealizations permit impossibly complicated calculations to be worked out approximately. The clever part of this procedure is that in many instances, the approximate results are close enough to reality to be useful; we gain a great deal of insight about the real forces that act upon the structure.

Our first idealization is that the members are perfectly rigid. That is, they will not bend or shear, stretch or compress, or snap in two, no matter how large the forces that act upon them. Second, we assume that the members are joined together at their ends by hinges that are perfect. A perfect hinge is infinitely strong. Adjoined members will not come apart at a hinge (known to structural engineers as a *pin*) no matter how large a force tries to separate them, and yet the pin exhibits zero friction. Third, we assume for ease of calculation that an external force (a *load*, recall) can act only at the joints of a structure rather than, say, midway along a member.

These three assumptions may seem to be extreme and unrealistic, but they are approximately true. Thus, real metal bars are not infinitely strong, but they are very strong in tension and compression, as we saw in chapter 1. Because pins are frictionless, and forces act only at joints, we can ignore shear and bending forces—the truss members experience only tension or compression, as we will see. A real force may act midway along a bar (for example, a weight suspended from the middle of the bar), but we can approximate this weight by two loads, each half as heavy, hanging from both ends of a member.

A few examples will serve to convince doubters among you about the efficacy of truss analysis based upon these idealizations. First, consider figure 2.1. Here we have four members connected end-to-end, with two of

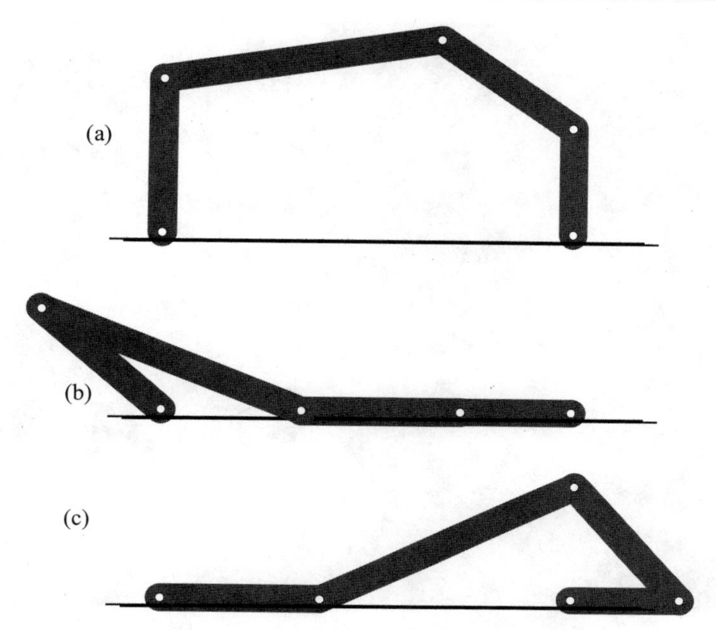

FIGURE 2.1. An unstable truss arrangement (a), consisting of four rigid bars, or members, joined by frictionless hinges or pins. The end pins are assumed to be fixed to the ground. Gravity will cause this four-element truss to collapse in one way (b) or another (c).

the pins placed upon the ground. The erected structure is unstable; under the action of gravity it collapses, as shown in the figure. This is realistic: we would expect real bars connected by real hinges to behave in this way, unless the hinges were very stiff. By making the truss pins frictionless we are bringing to the fore any instability that exists in the structure: it is free to fall down if it can and will not be saved from falling by the pins. In other words, there are no internal forces (within the truss) that prevent its collapse—only external forces can do that. Real joints connecting real truss members (fig. 2.2) are more resistant than these idealized pins; in this sense our truss analysis will be conservative—it will underestimate the stability of a truss. The underestimation can be regarded as part of the safety factor that engineers build into their calculations of large structure stability.

Common sense, and figure 2.1, tell us that only triangular combinations of truss members are rigid. Figure 2.3 shows how to make the structure shown in figure 2.1 stable: by adding extra members to make rigid

FIGURE 2.2. Wooden roof truss joints. (a) Metal plate and bolts. Photo by author. (b) Simpson tie. Thanks to Clayton Lewis for this image.

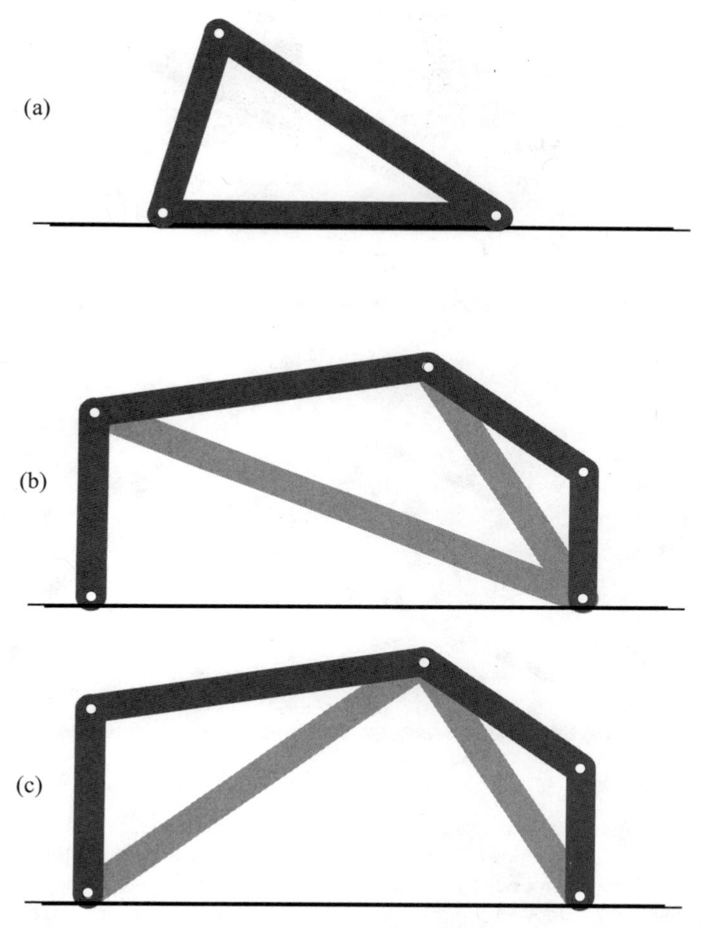

FIGURE 2.3. Triangular arrangements of truss members (a) are rigid. To make the structure of fig.2.1 rigid, we need to add extra elements so as to form triangles. Two possible schemes are shown in (b) and (c).

triangles. Again, our assumption of frictionless pin connectors is forcing us to think clearly about how to stabilize a structure composed of interconnected members. You may argue that the truss I have constructed in figure 2.3b is not really stable because one of the members is not part of a triangle. For example, a shed frame made in this shape would not be very good because the left support could be kicked away at the bottom, causing the shed to collapse. True, but there is one additional assumption about

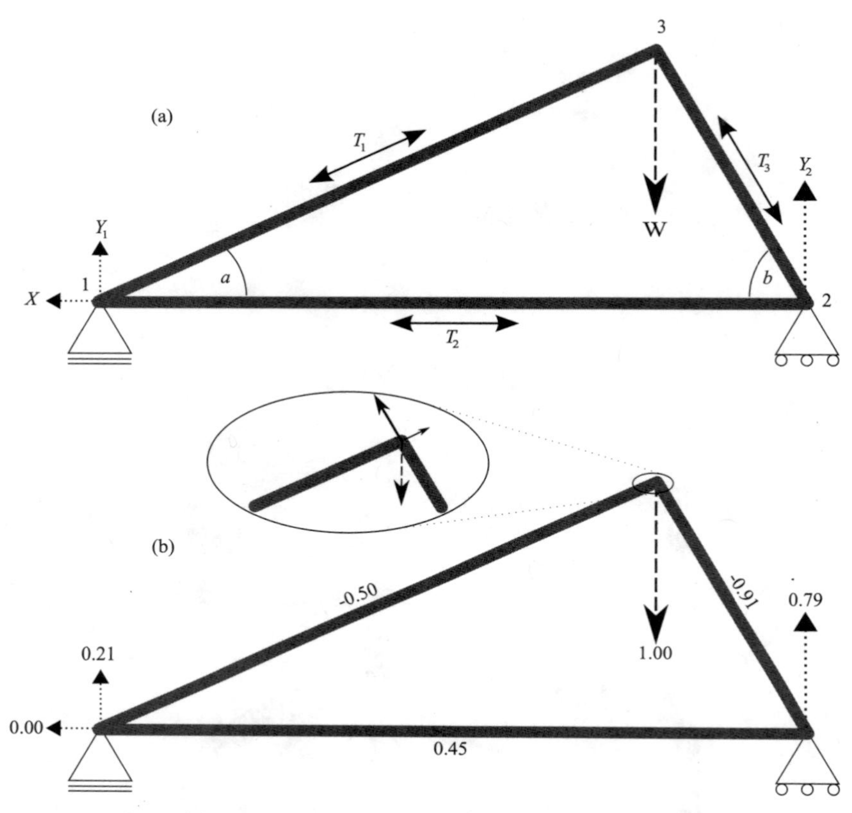

FIGURE 2.4. Ground forces that act upon a truss of three members. (a) The ground exerts reaction forces (dotted lines) that prevent the truss from moving left or right, up or down. The left ground force is assumed to hold pin 1 tight, while the right ground force allows pin 2 to move left or right, but not up or down. Tensile or compressive forces act along each element (two-headed arrows), and here we have also added an external load, W, acting on pin 3. Truss analysis leads to the forces shown in (b), assuming a load force of $W = 1.00$. The total vertical force of the ground exactly matches the load force ($0.79 + 0.21 = 1.00$), as we would expect. The horizontal ground force is zero, since there is no horizontal external force to match. An element is in tension when the force acting along it is positive and in compression when the force is negative. So in this case the bottom element is in tension while the top two are being compressed, which is what we expect for this particular configuration. The inset shows the forces acting at pin 3: the forces at any pin add up to zero.

truss theory that I have not yet mentioned. We have to say something about the manner in which pins on the ground are fixed to the ground.

To illustrate why, and how, we specify ground forces, consider figure 2.4. Here, we have a triangular frame set on the ground. The funny-looking triangles beneath pins 1 and 2 show us that these pins are on the ground. (If, as I do, you like tangible models to help fix an idea in your brain, think of the truss as a wooden building frame and the triangles as concrete piers.) Now, buildings and other structures do not fall through the ground and down to the center of the earth. At least, none that I have ever built did so. We have Sir Isaac Newton to thank for the explanation, of course, and his ideas figure prominently in the theory of trusses, as well as the theory of gravitation. Sir Isaac tells us that the ground exerts a vertical force that exactly opposes the downward force of gravity. In the figure I have assumed that the members are essentially weightless[1] and that the only heavy object that we need bother about is an external load, *W*, acting at pin 3. The ground forces must resist *W* and push back on the truss to prevent its falling down to the center of the earth.

So much for the vertical force exerted by the ground on those pins touching it. What about horizontal forces? We assume that the truss is stationary, and so the ground must exert a force to oppose any external horizontal force that acts on the truss. There is no such horizontal force in the case of figure 2.4; thus, the ground reaction force has zero horizontal component (pin 1). Later, we will see examples where there are horizontal loads acting on trusses (wind loading of roof trusses), and in these cases the horizontal ground reaction force is not zero. Note that only pin 1 is fixed firmly to the ground. Pin 2 is allowed to move horizontally along the ground but not vertically. We denote this by the "rollers" attached to the triangle beneath pin 2. Why do we treat the two ground pins differently? It turns out to be convenient for calculations, as we will see later, but there is also a practical reason. If the truss represents a metal bridge, then we might really choose to pin one end of the bridge firmly to the ground while letting the other end roll, to allow for thermal expansion. If both ends of the bridge were fixed then the bridge deck might buckle on a hot day, and we saw in chapter 1 how a small expansion can lead to significant buckling.

1. This assumption of massless truss members is commonly made because the external forces that act upon trusses are often much larger than the member weights, and so the member weights can be ignored. If we decide that member weight is significant, then we can include it by adding these weights as loads at each pin.

You will be forgiven for thinking that this chapter (so far) and chapter 1 are far removed from the soaring structures of the title. This is because I have had to, figuratively speaking, make bricks (beams and truss members) before I can build a house. Now that the bricks have been made, we can see how they are put together to build much bigger things. For the structural engineer, a key component of this building process is understanding the forces that act upon the individual members of a structure. Now that all the preliminaries are out of the way, we can calculate the forces that act along the truss members. We do so by applying Newton's laws. The essence of truss calculations is this: *The horizontal and vertical components of all the forces at all the pins must add up to zero*, so that the structure is stable. (See the inset of fig. 2.4b.) Note that in figure 2.4 we have six unknown forces to find: the three ground forces (horizontal and vertical forces at pin 1 and vertical force at pin 2) plus the force that acts along the length of each member. Sir Isaac gives us six equations (two for each pin joint, corresponding to horizontal and vertical components of force). This is good: with six equations we can solve for each of the six unknown forces. The results are shown in figure 2.4b. We see that, in this case, about four-fifths of the weight is carried by pin 2; the member 12 is under tension while members 13 and 23 are under compression. Changing the truss element lengths will change the angles *a*,*b* of the triangle, which will change the magnitude of the forces. The distribution of forces tells us how strong we need to make each member and whether the members need to be able to withstand tensile forces or compressive forces.

The elementary truss calculations for figure 2.4 are worked out in detail for you in note 3 of the appendix. If you are so inclined, you can sift through this example in detail to learn how to carry out truss analysis for yourself. If you would rather hit yourself over the head with a baseball bat than do math, then please at least glance at the appendix; it will show you how engineers can gain an appreciation of the forces that act upon a structure by approximating it as a truss. This is a trick that I will adopt several times in later chapters.

Bridge for Beginners

So much for the introduction to truss analysis; now let us put it to use. In figure 2.5 you can see a truss network that forms a bridge. Trusses are frequently used in this way because they are light for their strength. (Indeed, in figure 2.5 some of the large truss members are themselves trusses,

FIGURE 2.5. A truss bridge. Image from Wikipedia (Harvey Henkelman).

formed from smaller members.) Gantries and cantilever beams such as crane arms are also often built as trusses. The fact that the analysis is easy (at least, compared with that of solid beams with shear forces and bending moments, etc.) also helps. Bridges have point loads—such as a vehicle, or you—which are easily modeled in the truss analysis. In figure 2.6 we have different arrangements of what are called Warren trusses. You can see that, for the same load, the member tensions and compressions are higher in long bridges than in shorter bridges, which makes sense. The truss members along the top and bottom of each bridge are known as *chords*; the upper chords are in compression, whereas the lower chords are in tension. This also makes sense, if you stare at the figure and consider the load at the center causing the bridge to sag; the load will induce tensions and stresses within the truss network just as shown. (You may also recall the discussion of bending beams in chapter 1, with exactly this distribution of tension and compression.)

The benefit of truss analysis is that it permits us to quantify the forces of each and every member. The overall picture is of a beam made out of truss members with a bending moment applied by the load, causing the beam to

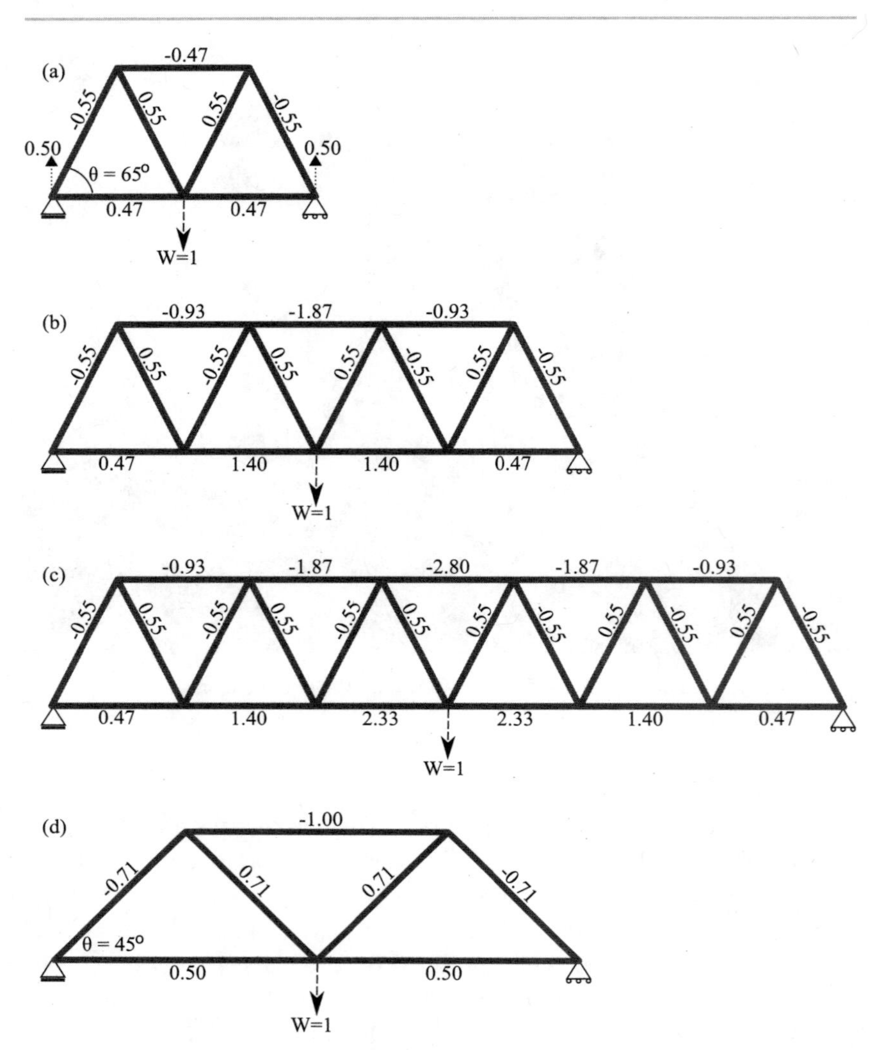

FIGURE 2.6. Warren truss bridges. (a) A Warren truss bridge with a unit load at the center. Weight is distributed evenly for this symmetric configuration; thus, the ground reaction forces are both 0.50. (b) A longer bridge has the same ground forces (not shown) but with greater tension and compression forces along the upper and lower chords. (c) Again in spades for an even longer bridge. (d) For a smaller element angle of θ = 45° the forces change. Compare this truss with truss (b), which is the same length.

be stretched along the lower surface and compressed along the upper surface—as in chapter 1, but now quantified. Forces are greatest along the chords. Note that the magnitude of forces acting on the internal members (called the *web*) is unchanging. Alternate diagonal members are in tension or compression (except for the central members near the load—note the symmetry of the configurations), but the magnitudes of these are generally less than the chord tension and compression. This is reminiscent of the neutral layer that lies along the center of a bending beam, which we discussed in chapter 1.

Truss analysis helps the structural engineer decide how strong each member needs to be so that the bridge will hold its design load. For the case of figure 2.6 we have placed the load at the most demanding point—in the middle—and have calculated the tensions and compressions in each truss member. We can see in figure 2.6c, for example, that the central member of the horizontal upper chord needs to be three times as strong in compression as the outer members. We can also see how changing the truss angle θ (fig. 2.6d) alters the forces. Changing θ will also alter the amount of material used to construct the bridge, and so the bridge designer will be able to optimize this angle. He would like to make the bridge as light as possible, to cut down costs and make construction more straightforward, but of course he needs to ensure that each member is strong enough to hold up under the expected forces. Fewer members means increased force per member, which requires stronger (and therefore heavier and more costly) members. So, fewer members may or may not be a good thing, depending upon the design load for our Warren truss bridge. Many applications seem to have settled for a truss angle of $\theta = 50°–60°$, and so we can assume that, for the strength and weight of steel tubing or bars commonly used as truss members, this range of angles leads to minimum cost for the required bridge strength. Some applications of trusses can be seen in figure 2.7.

What if the load is not in the center? In this case the distribution of forces is no longer symmetric, as you can see in figure 2.8. Of course, the design engineer must consider all possible load distributions before she makes up her mind about the bridge design. I will investigate other truss bridge designs in the next section, where we will see how the forces change as a load (a car, say) crosses the bridge. I will finish this section by bringing to your attention another choice that the designer must make for her bridge: should the deck (the road pavement) be along the upper chord or the lower chord? Both these cases are illustrated in figure 2.8. A through

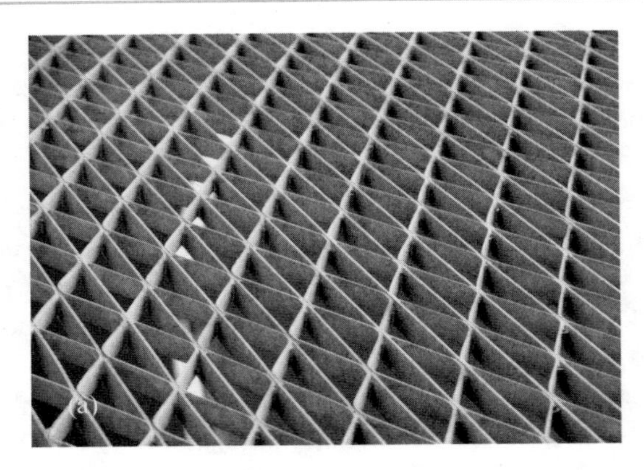

FIGURE 2.7. Various uses for trusses. (a) This walkway is light and rigid because of the truss construction. I thank Kees van Hest for this image. (b) Truss sign post, truss tower, and truss tower supports. Photo by author. Trusses are everywhere. Perhaps we can also regard the wheel (c) as a truss construction.

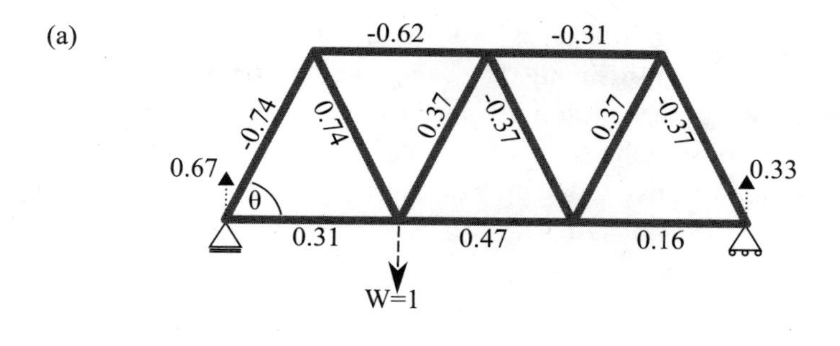

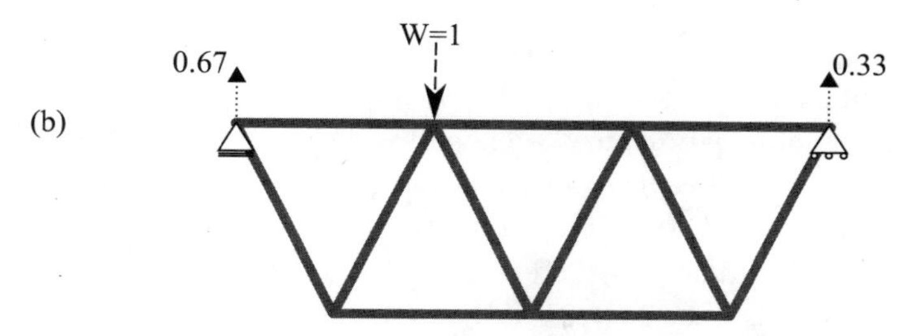

FIGURE 2.8. Asymmetric loading. (a) A Warren truss bridge with an off-center unit load. Now the force transmitted along the members is not symmetric. This is a through design, with the bridge deck along the lower chord. The forces shown correspond to a bridge angle of $\theta = 65°$. (b) The same Warren truss, but this time as a deck or underslung bridge. The deck is along the upper chord. The member forces are reversed: tensions become compressions, and vice versa.

design has the deck and the vehicle load along the lower chord,[2] whereas a deck or underslung design places it along the upper chord. For both types of bridges the upper chord is in compression and the lower in tension.

It turns out that for the case of a uniform load, the central web members are *zero-force* members: they do nothing.[3] They are freeloaders, taking up none of the burden shared by the other members. In principle they could be removed, and the bridge would still be structurally sound, but only for this load distribution. For other load distributions these central members are not freeloaders (there are no zero-force members for the load distribution of fig. 2.8, for example). For some bridge designs certain members are

2. The road surface—the deck—may be a significant fraction of the total bridge load.

3. A uniform load is evenly distributed over the joints, except that the end joints are assigned half the load of the other joints because they are loaded from only one side.

always freeloading. In that case, why bother to have them? Remember that truss analysis is approximate. The bars, beams, or tubes that constitute real truss members are not infinitely strong. Long members are subject to bending forces and may buckle under a heavy load; in such cases a zero-force member may be inserted to provide extra rigidity and prevent buckling.

Determinate or Indeterminate?

The number of forces that we must calculate (which, for the trusses we have looked at so far, is the same as the number of equations) increases with bridge size. So, for the bridges in figure 2.6a–c we solved 10, 18, and 26 equations for 10, 18, and 26 unknown forces. For large truss structures the number of equations that we need to solve in order to determine the force distribution becomes very large indeed. Structural engineers have developed analytical techniques that help with such calculations,[4] but as the number of equations increases, even these techniques become unwieldy; and it is necessary to turn to a computer to churn through the equations. For realistic three-dimensional structures the problem is exacerbated because each joint has three components of force associated with it (and so three equations) rather than two (fig. 2.9).

In fact for many truss structures, the problem is worse than I have indicated. For some trusses the number of equations that emerge from the structural arrangement is less than the number of forces that we need to calculate. In these cases it is impossible to calculate all the forces. For all the trusses that we have looked at so far (shown in figs. 2.3, 2.4, 2.6, and 2.8), the number of equations equals the number of forces, and so the forces can be calculated. Such systems are called *statically determinate*. They are characterized by the following statement: *The forces that act at all points on a statically determinate structure are completely determined by the external forces.* Think about the trusses that we have looked at so far. All of them have ground reaction forces and loads, both of which are external forces. Specifying these external forces has enabled us to determine the forces that act internally at every pin joint of the truss structure. Indeed, because of our simplifying, idealized assumption of zero pin friction and infinitely stiff and strong members, we have banished the need to consider internal forces. The result is a calculation scheme—truss analysis

4. The so-called "method of joints" and "method of sections," plus vector notation.

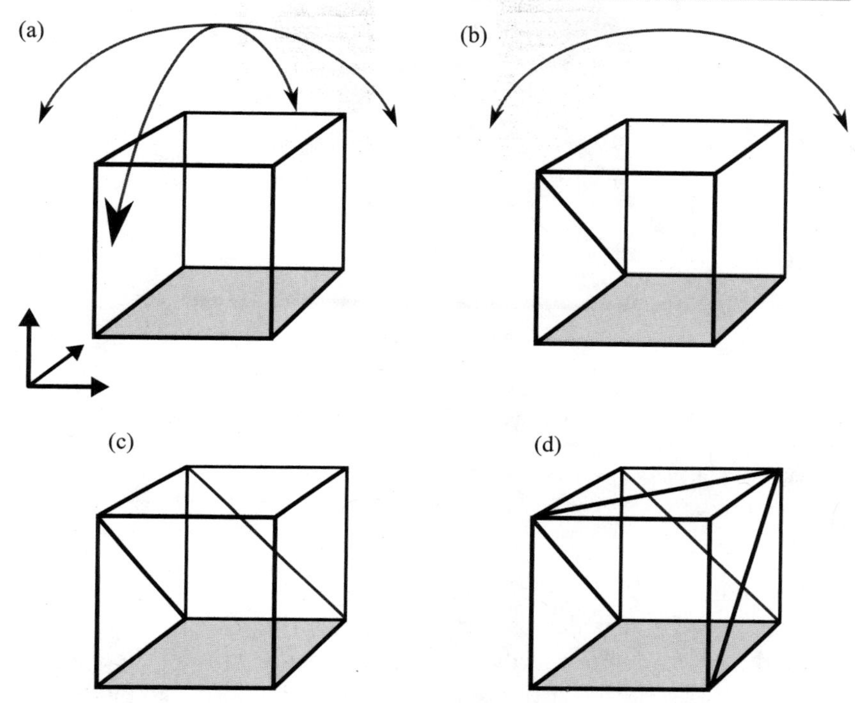

FIGURE 2.9. Stability in 3-D trusses. (a) A 3-D truss made up from six members in the form of a cube is unstable: it can collapse by falling in any direction and can also collapse by twisting. Three directions, and therefore three components of force, are associated with each pin joint. (b) With one diagonal brace, one face becomes two triangles and so is rigid in its plane. Now the cube can collapse in only two directions. (c) With two diagonal members (not in the same plane) the cube will not collapse. (d) For a real truss, extra diagonal members will increase rigidity.

—that enables us to determine the effects of external forces at internal points of a truss.

When the number of equations we can extract from a structure is less than the number of forces—a situation that arises when internal forces come into consideration—the structure is said to be *statically indeterminate*. Truss analysis is not sufficient for such structures; it cannot calculate all of the forces that are significant. A simple example of a statically indeterminate truss structure is shown in figure 2.10. More complicated and realistic examples involve structures for which internal forces are significant. Beams—real beams like those of chapter 1 that can bend and twist

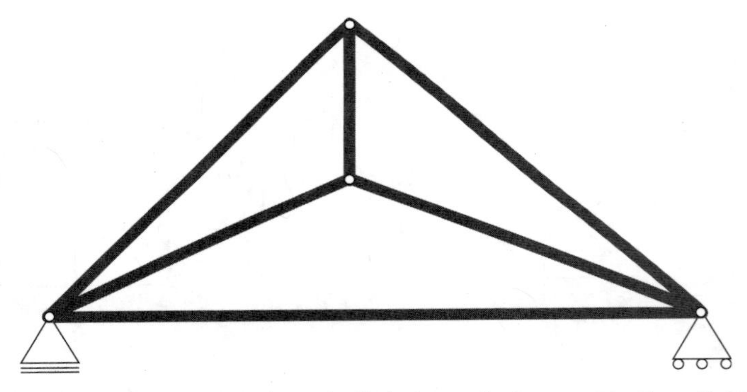

FIGURE 2.10. This truss structure is statically indeterminate: we cannot use the truss analysis methods that have served us well up to now because for this arrangement (and many others) the number of forces that need to be determined (nine) exceeds the number of equations we have for the structure (eight).

and stretch (not idealized truss members)—are common structures that often have to be regarded as statically indeterminate because to understand them we need to specify their bending moments and shear forces, etc., which are characteristics representing internal forces acting *between* molecular components of the beam and external forces acting *upon* molecular components of the beam. Recall that in chapter 1 we noted some simple observations about beam theory without resorting to detailed calculation. I ducked the math because beam theory is much harder than truss theory—and the reason for that is static indeterminacy.[5] Beam theory is a way of summarizing the effects of myriad internal forces (forces that our simple truss analysis can't handle) with a few key parameters and concepts —but the calculations are difficult.

To provide you with a simple concrete example that illustrates the basic concept of static determinacy and indeterminacy, look at the pub signs shown in figure 2.11. The figure caption explains why one sign is statically determinate and the other is not.

5. The mathematical definition of a stable truss is one for which $m + r \geq 2j$. That is, the number of members, m, plus the number of ground reaction forces, r (3, usually), must equal or exceed twice the number of joints, j. Equality occurs for statically determinate structures. If $m + r$ is greater than $2j$, then the structure is stable but statically indeterminate; if $m + r$ is less than $2j$, the structure will fall down. Now that you know the mathematical definition, check the structure in fig. 2.10 to confirm that it is statically indeterminate.

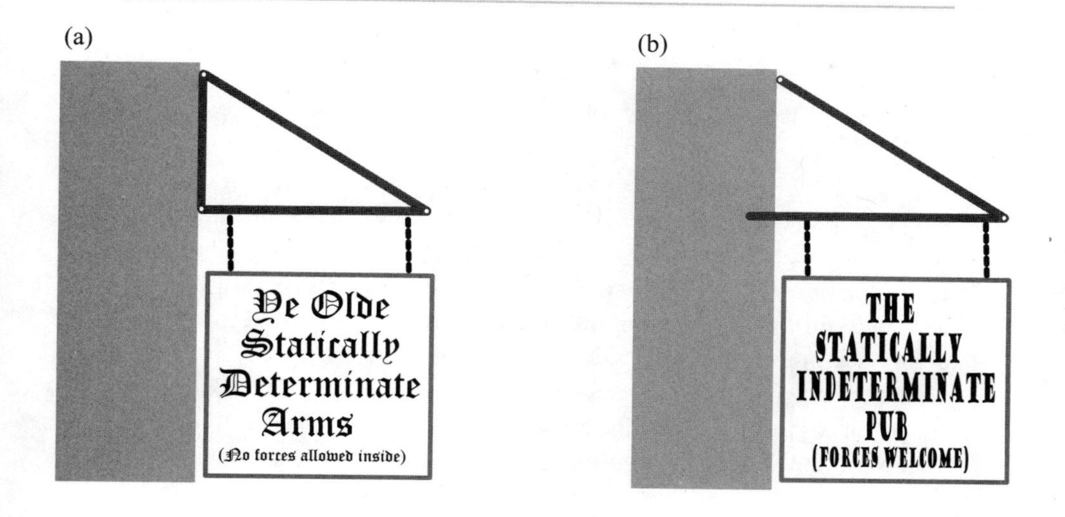

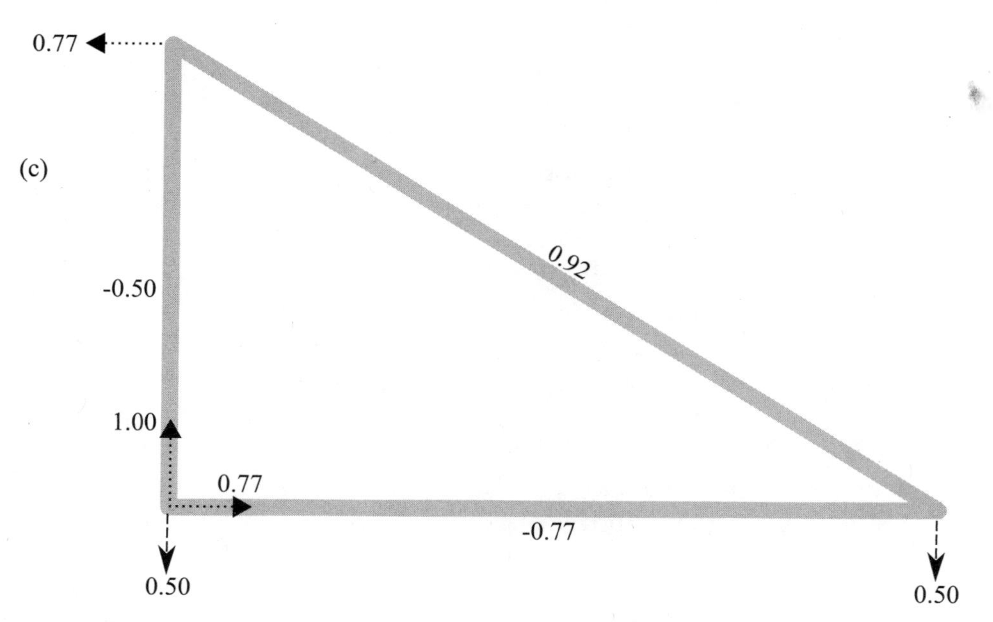

FIGURE 2.11. (a) The triangular truss is a statically determined structure. (b) If the horizontal member is a beam embedded in the wall rather than pinned to it, the structure is statically indeterminate. We cannot evaluate, say, the tension in the diagonal member without knowing more about the internal characteristics of the horizontal member, such as how stiff it is. (c) Here, the forces for (a) have been calculated assuming that the pub sign has a weight of 1.00, evenly divided between the two lower joints. The horizontal and vertical members are in compression, while the diagonal member is in tension. The wall exerts three forces (dotted lines) that hold the sign in place.

Pratt, Howe, and Influence Lines

Many different types of truss bridges have been tried over the years. Normally, trusses are used for bridges with a small span (40–500 meters, or 130–1,650 feet); we will see in chapter 5 how much larger spans are attained. Many truss designs originated with the expanding railways of the nineteenth century, particularly in North America, as thousands of miles of new track were laid down over difficult terrain. The Pratt truss bridge, a statically determinate structure, is analyzed in figure 2.12 for a load that moves across the bridge. For such a load, the forces that act upon a particular member change with time; the graph of these changing forces is known as the member's *influence line*. The influence line for one of the diagonal members is plotted. Note from figure 2.12a–c that the upper chord is always in compression and the lower chord always in tension. The diagonal members can be in either compression or tension, but in fact for longer bridges with more diagonal members, only the central two can be in compression; the other diagonal members are always in tension. This is good news for the bridge designer, because light, strong, and inexpensive steel cables can be used for members that are always in tension. A member that is subjected to compression must be a sturdier and more expensive piece of equipment. The diagonal members are longer than the vertical ones; if we find that they are in compression, the extra length again means extra weight and expense. For a Pratt truss bridge, only the two central diagonal web members need to be so sturdy; cables or light beams can be used for the others.

Look what happens if we make the seemingly innocuous change of switching the slope of a diagonal member. We see in figure 2.13 that the forces acting upon a diagonal member are large, compared with the same member of a Pratt truss bridge. Also, the lower chord is no longer always in tension, and so its members will have to be robust enough to withstand compressive forces. Not such a good design.

There is a lot more to bridge design than I have indicated in this introduction to the subject. In the first place, bridges are three-dimensional structures, and so, as we saw with the cube truss of figure 2.9, there are more ways for them to collapse. A bridge may consist of two parallel trusses with the bridge deck between them. This design is known as a *pony*: it is uncovered (open at the top). A covered bridge, in truss structural terms,[6]

6. As distinct from the old-fashioned *wooden* covered bridges that look like sheds across short spans.

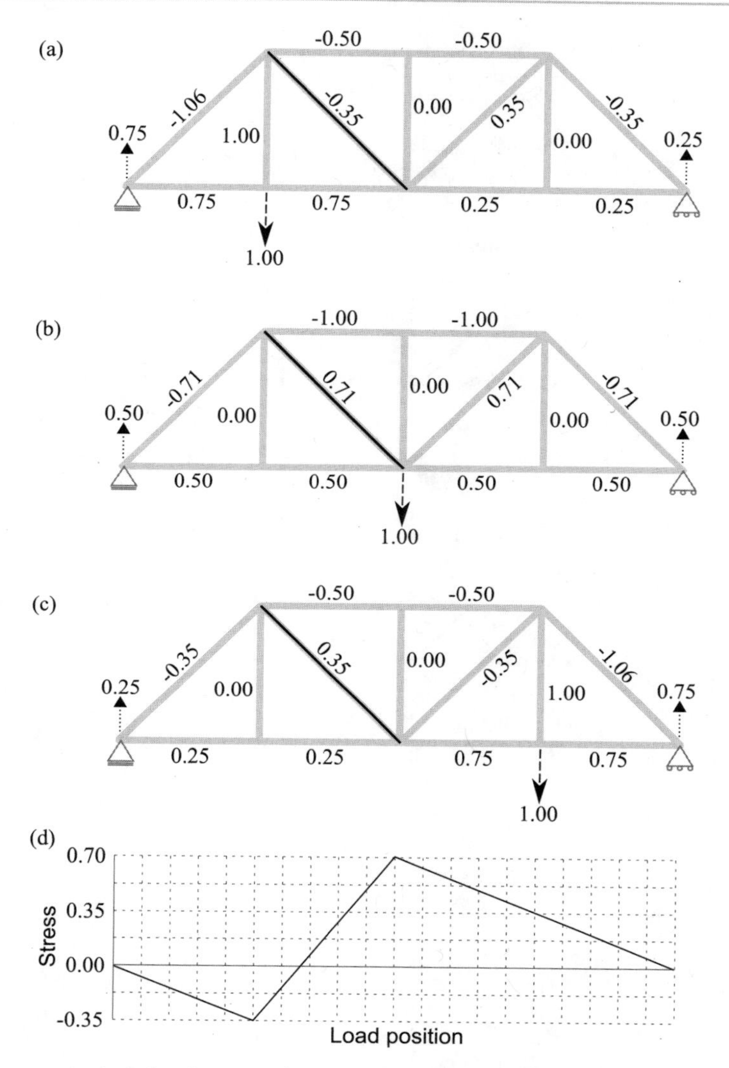

FIGURE 2.12. Analysis for the statically determinate Pratt truss bridge, with a unit load crossing at the left (a), the center (b), and the right (c). The chart in (d) shows the influence line for the black diagonal member of drawings (a)-(c). Longer Pratt trusses have the same pattern repeated, with diagonal trusses angled down towards the center. This is an efficient design.

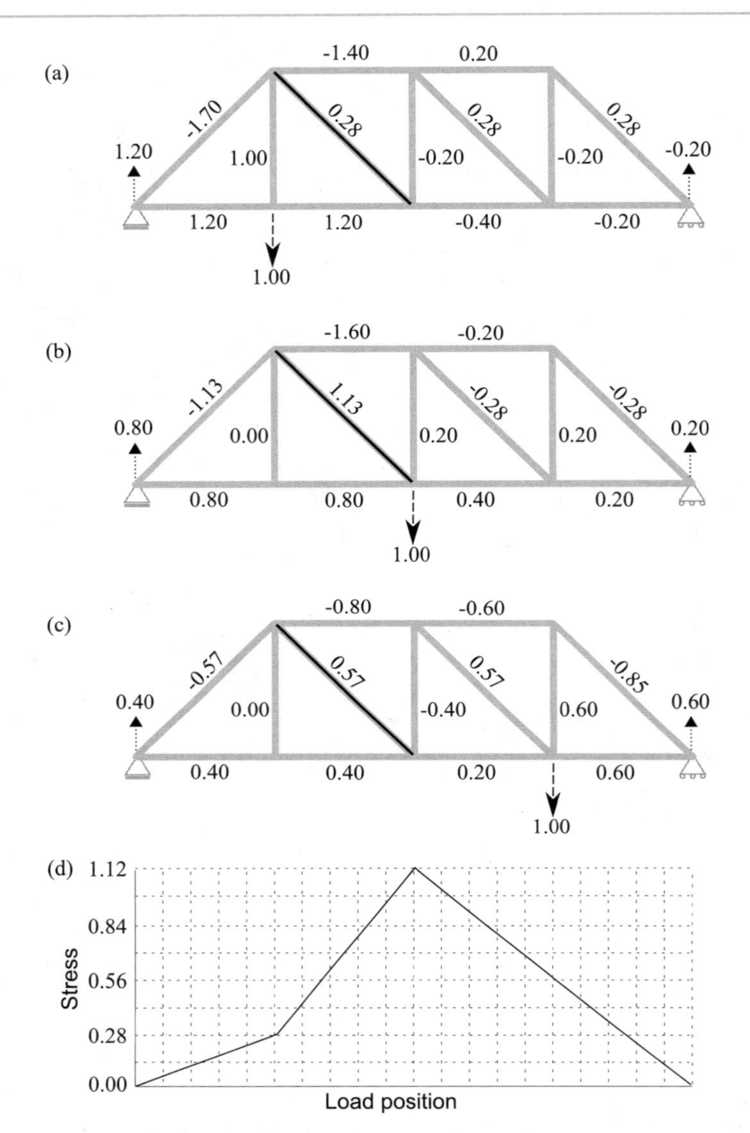

FIGURE 2.13. A truss structure similar to that of Pratt but not so good. As the load moves across the bridge, as shown in (a)-(c), some upper-chord members are under tension and then under compression. They have to be strong for both types of force. The influence line (d) shows larger forces than for the Pratt truss.

is covered over the top with another truss, which adds extra rigidity to the bridge.

Secondly, bridges are not just subjected to static loads. When the load is moving, the analysis becomes much more complicated. The *dynamic loading* of a bridge introduced surprises to bridge engineers when truss bridges were first built. The main advantage of truss design (apart from the ease of analysis, at least for statically determinate structures) is the ease of construction, low cost, and light weight. Traditionally, as we will see in chapter 5, bridges were built of stone and, as a consequence, were much heavier than the traffic which crossed the bridge. Because of their weight, therefore, bridge designers in ancient times just needed to build bridges that would not fall down—they did not have to worry about the extra weight of traffic, since it was always so much less than that of the bridge. If the bridge stood upright under its own weight, then it certainly would not be toppled by the tiny extra weight of a cart or person passing over it.

All this changed with truss bridges in the nineteenth century. Lightweight truss bridges were introduced at just the time when the traffic load (from steam locomotives in particular) became very heavy. The combination of light bridge and heavy traffic meant that bridge engineers had to tackle a new problem. For the first time they had to take into consideration the dynamics of traffic passing over their bridges: the physics of bridge design was required to incorporate moving masses, since these could no longer be ignored. I will discuss this problem in more detail in chapter 5. Here, we must be satisfied with static analysis. Note that the influence lines that we have constructed in figures 2.12 and 2.13 do not constitute a dynamical study; we have simply looked at the statics of bridge loading at different times, or rather, at static loads at different positions on the bridge. The dynamics of a moving load suddenly impacting the bridge lead to problems that were unexpected. To be continued.

Ponte Pasta

I cannot leave the subject of truss bridges without at least mentioning the growing field of spaghetti bridge building. No, this is not a misprint, but please do not worry the next time you drive over a truss bridge. You do not need to take your eyes off the road to check if the structure is of steel or spaghetti, of timber or tagliatelle. Spaghetti bridge competitions are a popular and educational adjunct to many a university engineering degree program (fig. 2.14), and we can now appreciate why. Uncooked spaghetti is

FIGURE 2.14. A spaghetti bridge. Thanks to Michael Karweit of Johns Hopkins University for permission to reproduce this image.

rod shaped and structurally quite weak. It snaps easily when bent, and so a budding bridge builder can use it only for structures requiring *axial forces* (forces that act along the direction of the member). It does not stretch much.[7] However, when spaghetti is glued together in truss shapes, bridge structures can be made (cooked up?) that support quite respectable loads. This practical exercise teaches students about truss design and emphasizes how the strength of trusses resides in the overall structural design rather than in the individual members.

Johns Hopkins University, in Baltimore, has run an annual spaghetti bridge competition for over a decade, and the rules are quite strict. The structure must have a span of at least a meter, with additional dimensional restrictions: the height is limited and the deck must permit a load of specified size to pass along it. Bridge weight is limited to 750 grams (that's 1 pound 10 ounces dry weight—serves six people), and the winning design is the bridge that supports the greatest load. One recent winner held up an

7. You will not need to strain this spaghetti. Technical pun—apologies.

impressive 56-kilogram (123-pound) load. Okanagan College in British Columbia, Canada, holds an annual spaghetti bridge building contest that attracts entrants from around the world. Its 22nd winner in 2005 was Andras Koves of Budapest Polytechnic in Hungary; his spaghetti bridge weighed less than 1 kilogram and yet supported a load of 257.33 kilograms (566 pounds).

Raising the Roof

Roof trusses have been a part of timber-framed buildings since classical antiquity. In ancient times they were usually associated with prestigious, high-end buildings such as churches or mansion houses. This was probably because only large buildings required trusses: smaller dwellings could be roofed over with simple beams. Many variants of roof trusses survive from medieval times in Europe, and the variety of truss structures is large (see figs. 2.15–2.17 for some examples).

Figures 2.15–2.16 indicate something of the evolution of European timber roofs during the Middle Ages. The relatively simple crown post roofs of the thirteenth and fourteenth centuries (both of these illustrations show a *king-post* roof; the *queen post* has two vertical braces) gave way to the ornate *hammer beam roof*, which looked better and opened up more space. Hammer beams were also able to span a wider space. Note the similarity of the truss arrangement with that of later truss *girders*, such as we saw in the diagonal truss elements of the bridge in figure 2.5. Hammer beams are still popular today, though they peaked (as it were) in the late fourteenth and fifteenth centuries. Arch braces succeeded hammer beams; these were more difficult to construct, but were able to span still larger spaces and opened up the roof space even more.

Figure 2.15 provides only a very broad-brush indication of the way in which timber roofs evolved. There were dozens of variants, each well-adapted to a particular circumstance. Generally, however, the trend was for timber roofs to evolve from cluttered to open designs, from small to large spans, and from functional to functional-and-graceful. I will here divide the menagerie of pitched roof trusses into two types: those specimens with a horizontal lower chord (a *tie beam*) and those with a raised lower chord. This division is not architectural or historical, but it suits my purposes. Some of the flat-lower-chord trusses are illustrated in figure 2.18; while this selection is not exhaustive, it does get across the kind of variations that are practical. We will soon apply what we have learned about truss analysis

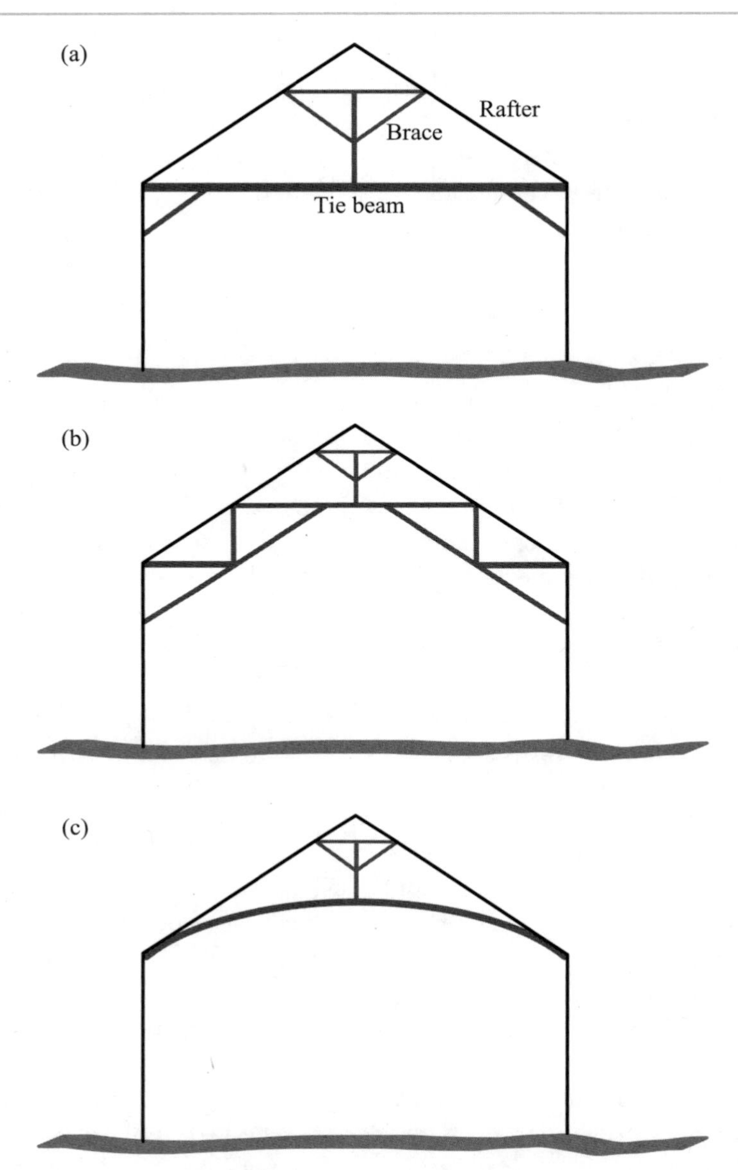

FIGURE 2.15. The evolution of roof truss styles during the Middle Ages: (a) king-post roof; (b) hammer beam roof; (c) arch brace roof.

FIGURE 2.16. Many and varied wooden roof trusses survive from important or high-status buildings of medieval Europe. Here is a king-post truss roof from Old Romney Church in Kent, England.

FIGURE 2.17. A modern timber arch beam. I thank Keith Walker for this photo.

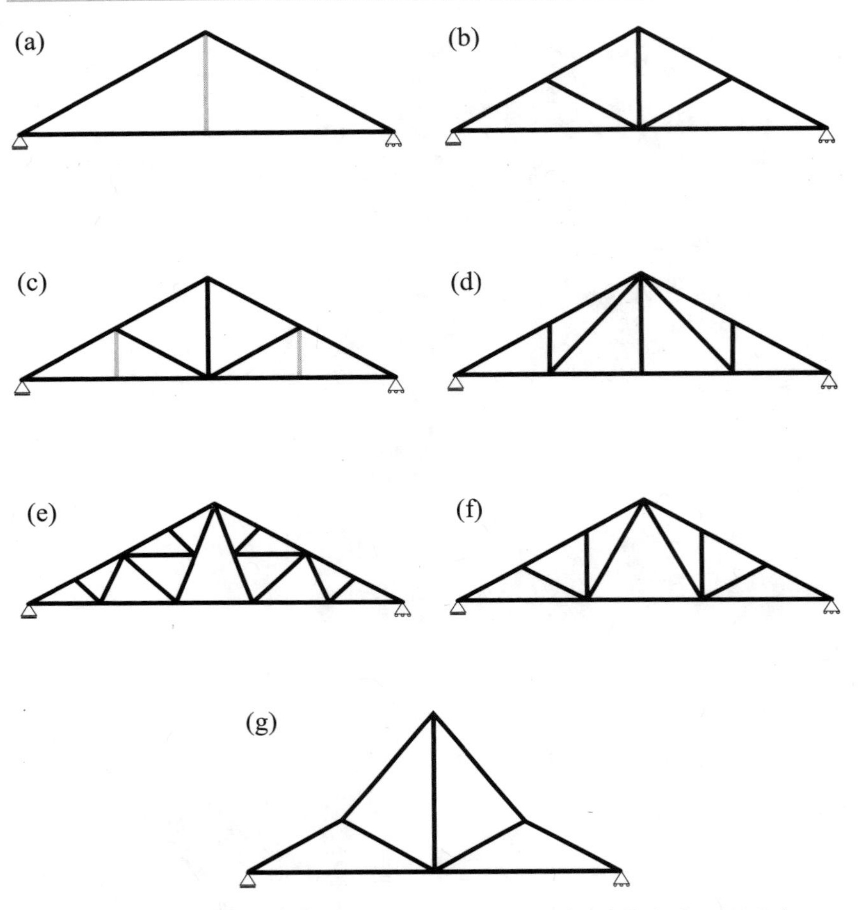

FIGURE 2.18. Pitched roof trusses with a horizontal lower chord. Zero-force members are shown in gray. (a) King post; (b) Palladian; (c) Howe; (d) Pratt; (e) Fink; (f) Fan; (g) Bell.

to a few representative designs so that we can gain some insight into why they arose and why there is not a single, best roof truss design that has become universal. One feature of the flat-lower-chord trusses is that the ceiling may be placed below the trusses, so that they are hidden from view, or placed above the trusses, exposing them as an architectural feature. Some trusses are attractive and worth exposing, whereas others are a messy framework that is best kept out of view.

The second type of roof truss is characterized by a raised lower chord. I illustrate a few of these designs in figure 2.19. Raised lower chords mean more headroom, and a greater feeling of spaciousness within the building. Generally speaking, the stresses that act upon the truss members of this second type are higher than for the first type, and so raised-lower-chord trusses must be stronger than flat-lower-chord trusses covering the same span—they must be made either from a stronger material or with a larger cross section. Because they make greater demands upon the structural strength of the truss members, raised-lower-chord roofs are a more recent innovation.

The Polonceau truss illustrated in figure 2.19 was first used in the roof of

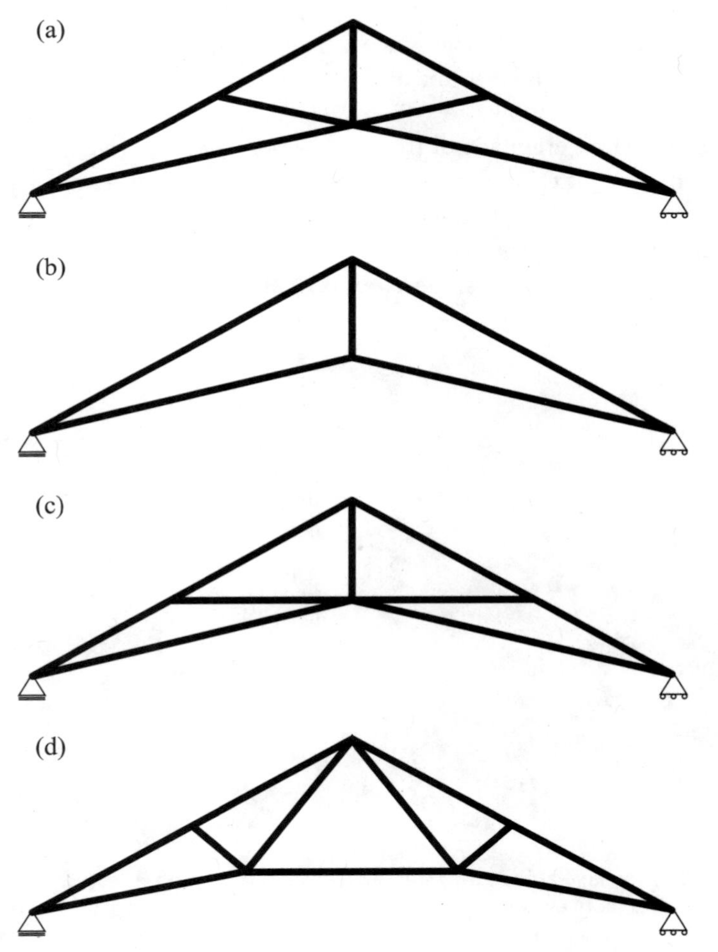

FIGURE 2.19. Pitched roof trusses with a raised lower chord: (a) scissors, with vertical web member; (b) Swiss; (c) German; (d) Polonceau.

the Paris-Versailles railway station in 1837. In fact, the rise of railways in the nineteenth century provided a great spur to roof truss development, at a time when a new building material—cast iron—was becoming plentiful. As we saw in chapter 1 cast iron is very strong in compression, like stone, but unlike stone it can be formed easily to any desired shape, and it proved to be a very versatile material that was widely applied to large-span roofs, particularly in the nineteenth century (see fig. 2.20). Railways required sheds to house the locomotives and platforms to cover the paying traveler from inclement weather. As stations grew in size, with many platforms, it became necessary to provide roofs with ever larger spans. Large spans induce large stresses, and so it was necessary to build light but strong roofs that were well designed to maximize the strengths of the truss material without risking roof collapse.

Earlier roofs were made of stone and wood and required the heavy tie beams, braces, rafters, and *purlins* (members running perpendicular to the rafters, along the length of the roof) seen in figure 2.16. The weight of these materials limited the spans that could be attained by stone and wood roofs. As we saw in chapter 1, stone is weak under tension, whereas wood is

FIGURE 2.20. The first large railway stations appeared in England. The many converging tracks and locomotives gave rise to large engine sheds and covered platforms for passengers and required the development of bigger roofs spanning much larger areas than had previously been necessary. Adapted from an 1852 illustration of King's Cross Station in the *Illustrated London News*.

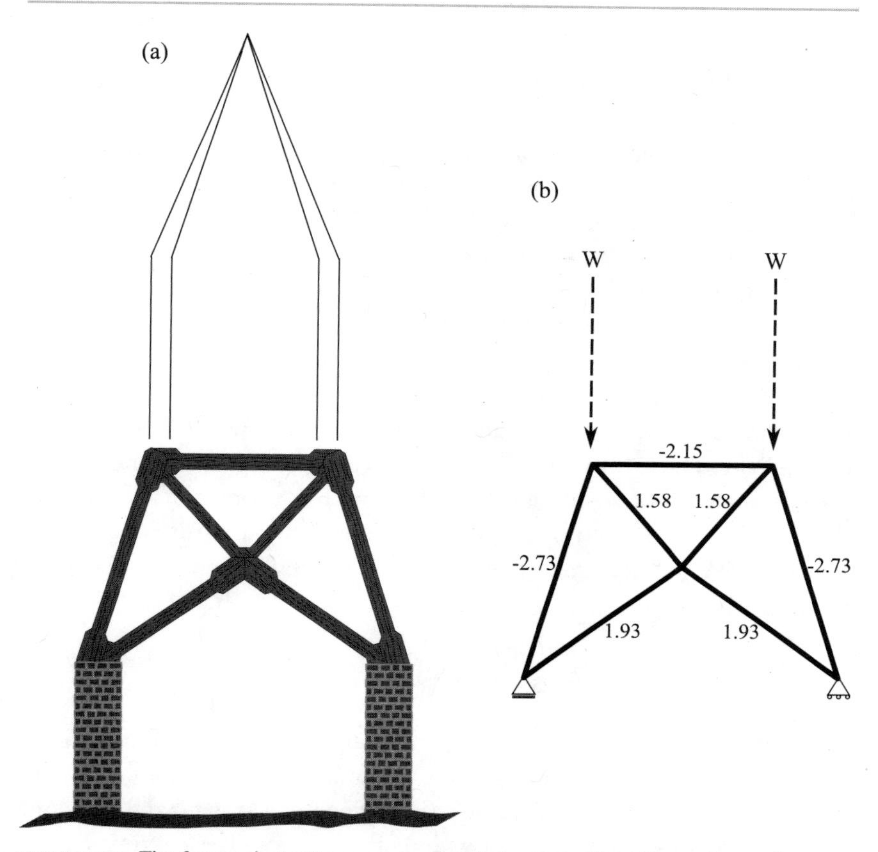

FIGURE 2.21. The forces that act upon a medieval church roof. (a) A wooden roof truss rests on stone walls and supports a steeple. (b) If the weight of the steeple is W on each of the upper chord joints, the stresses that act upon the truss members are as shown. A negative stress corresponds to compression and a positive stress to tension. Thus, for example, the horizontal member of the upper chord is subjected to a compressive force of $2.15W$. Since the upper chord is under compression, it might be constructed of stone. The lower chord and the web members are under tension and so should be made of wood.

strong. Therefore, in older roofs stone was used for those parts that are subjected to compressive forces and wood for those parts that are in tension.[8] Happily, the upper chord usually carries a compressive load and so

8. If the compressive load is not too large, wood can support it—my house and yours are still standing—but for very large compression forces, stone is the traditional material of choice. Under tension, stone is pathetically weak, as we saw earlier.

stone was used on this exposed part of the roof, whereas wood was used inside, safely out of the weather. This is why medieval roofs have lasted to the present day.

We can easily show, from our knowledge of trusses, that a sensible truss design will lead to the kind of force distribution that permits a stone upper chord and wooden lower chord. In figure 2.21 we have a church roof that supports a steeple. The steeple weight is transferred via the wooden trusses to the walls. The trusses have a raised lower chord, as church roofs often do, so that the simple truss design is open to view; the worshipers below can admire the lofty roof while contemplating higher things. Note from the figure that our simple roof truss does indeed call for compressive forces on the upper chord (read "stone" or "strong wood") and tension on the lower chord and the web members (read "wooden members"). So, the combination of wood and stone deals with the forces that arise from the truss structure, distributing the steeple weight, and this combination puts the weather-resistant stone on top, protecting the wooden members underneath.

Modern Roof Trusses

Those ancient builders made lasting and beautiful structures, but their understanding of the physics and engineering problems was entirely empirical —trial end error. The "five-minute rule" of chapter 1 applied and had done good service, in a rough-and-ready kind of way, since classical times. Later builders developed more of a theoretical appreciation of the forces that acted upon the structures they were building, as we will see in chapter 4. Today, we have a much better understanding of these forces and of the strengths of the building materials we use, as noted in chapter 1. Because of this understanding we can design roof structures and be confident, before any construction begins, that they will last a lot longer than five minutes. The humble wooden roof trusses that many of you see every day (particularly if you live in North America) are the result of much analysis and development. Here, I will again apply the physics of statically determinate truss structures to demonstrate the advantages and disadvantages of three common roof truss designs.

In figure 2.22 I have analyzed three of the roof truss types shown in figure 2.18, stripped down to the bare essentials for simplicity. Thus, the

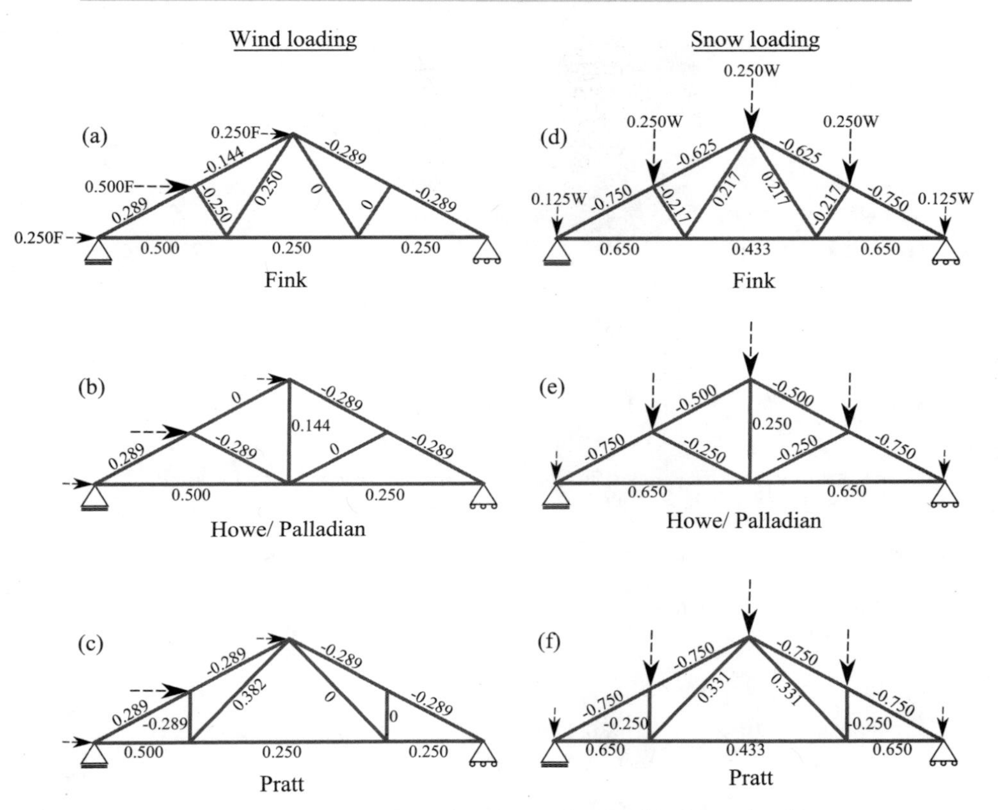

FIGURE 2.22. Analysis of three roof trusses subjected to wind and snow loading. (a)-(c) Fink, Howe / Palladian, and Pratt roof trusses subjected to wind loading of force *F*. The stresses on the truss members are shown (fraction of *F*). As always, negative stress indicated compression. (d)-(f) The same roof trusses subjected to snow loading with a total weight *W*.

Fink truss is the simplest of its ilk;[9] the Howe truss has been shorn of its zero-force members (and so becomes the "Palladian" truss of fig. 2.18), and so has the Pratt truss. Even with this simplification, figure 2.22 looks too busy to be easily digested. So, to summarize the results of the analysis, please consider table 2.1, in which the key data are presented in a more

9. You may have observed that the Polonceau truss of fig. 2.19 is simply a "raised Fink." Indeed, the differences between many of these different truss assemblies may seem to be minor, but they are significant. (If they were not significant, then mass production economics would have led to a standardized roof truss design.)

TABLE 2.1. Results of truss analysis for the Fink, Howe, and Pratt roof trusses of figure 2.22, for two pitch angles and two web member lengths

	Fink		Howe		Pratt	
	30°	45°	30°	45°	30°	45°
Wind tension	0.500	0.500	0.500	0.500	0.500	0.559
Wind compression	0.289	0.354	0.289	0.354	0.289	0.354
Snow tension	0.650	0.375	0.650	0.375	0.650	0.375
Snow compression	0.750	0.530	0.750	0.530	0.750	0.530
L	1.00	1.58	0.87	1.21	1.05	1.62
L_2	1.33	2.11	1.44	1.91	1.34	2.12

Note:
Snow tension / compression = maximum tension / compression within truss for a (vertical) snow loading, W; numbers represent tension / compression force as a fraction of the total snow load, W.
Wind tension / compression = maximum tension / compression for a (horizontal) wind load force, F, as a fraction of the total wind load.
L = length of web truss members, as a multiple of the roof width (lower chord length).
L_2 = same as L, but with web elements that are in compression doubled up.

palatable manner. In this table you will find not only the maximum tension and compression for both wind and snow loading, for each type of truss, but also the length, L, of webbing (the interior truss members) that is required for a given truss type. This is an economic consideration: if several trusses perform equally well, the best choice will be the one that requires the least amount of material to build. The table shows the web truss total length, and omits the upper and lower chord lengths, since these have to be there anyway for the type of roof that I am considering. The table shows these data for a 30° roof pitch angle, as illustrated in figure 2.22, and also for a steeper 45° angle.

The roof trusses in modern houses (see fig. 2.2) are usually made of wood (2 × 4s, typically, or 2 × 6s).[10] Wood, as we have seen, is stronger under tension than compression. For this reason building contractors may want to double up the web members that are subjected to compression—

10. For the uninitiated, I will explain the dimensions. This was a mystery to me when I first came to live in Canada, and there may be a few more like me out there, so here goes. The dimensions refer to the outside dimensions of the plank cross-section in inches before finishing. A finished plank is a half-inch less than the rough-cut version, and so, for example, a length of 2 × 4 has a cross section of 1½ × 3½ inches.

i.e., form the member from two planks of 2 × 4s rather than from a single plank. This increases the amount of material used to form the truss, and this length, L_2—the length of wood needed to form the truss web if members under compression are doubled—is also shown in table 2.1. In a real roof truss, extra zero-force members will be added for stiffening, but these are omitted here—the analysis is already beginning to give me a headache, so enough is enough. Table 2.1 and figure 2.22 show the key results, which I can summarize as follows:

— The maximum compression forces and the maximum tension forces are pretty similar for all three of these simple truss types. Unsurprisingly, wind load is heavier for the more steeply pitched roofs, whereas snow load is lighter.
— The lower chords are always in tension. For snow loading, the upper chords are always in compression. For wind loading part of the upper chord can be in tension.
— The Fink truss requires the least web material for low pitch angles, while the Howe truss wins for steeper angles.

The purpose of this exercise is to show you that we can apply the theory of trusses to real situations and draw conclusions about the efficacy of different designs. The real world is more complicated than my analysis suggests, of course, because real roof truss members are not perfectly stiff and are not connected by frictionless pins. A detailed analysis would add a few well-chosen zero-force members for rigidity and would take into account the strength (and cost) of joint connectors, be they screws, bolts, or metal plates. Despite the simplifications adopted here, I think it is clear how structural engineers go about their business of designing trusses—different ones for different situations.

Beams and Trusses

In chapter 1 we saw how the shape of a beam influences its stiffness in a particular direction. Thus, I-beams and corrugated metal sheets are resistant to bending about one axis; because they combine light weight and strength, they are good for roof construction. Trusses, we have seen here, are also lightweight and strong, though they have their limitations. Combinations of beams and trusses permit structural engineers to design an endless variety of roofs. Sometimes the variations are purely functional,

FIGURE 2.23. The roof of the Great Hall in Stirling Castle, Scotland. This heavy and intricate hammer beam truss is both functional and ornate. It opens extra space denied to a flat lower chord truss design and therefore lets in more light. I thank Bill Webb for this image.

and sometimes they exist mostly for aesthetic reasons. For me and many other people the beauty of roof design is a mixture of the two: functionality and aesthetics, science and art. I can appreciate the architectural appeal of exposed beams and trusses, such as the roof shown in figure 2.23—my mathematical brain finds symmetry attractive—but also I find beauty in the clever combination of roof truss members and beams. Its designers and craftsmen did their physics (empirically) and produced a structure that is both functional and aesthetically appealing.[11]

To illustrate how we might combine beams and trusses, consider figure 2.24. You can see that not all the joints form triangles, and so these roofs are not purely truss structures. Recall that trusses are subjected only to axial (tension and compression) forces, whereas beams may be subjected to bending, shear, and other forces. If all the joints were replaced by pins, the structures of figure 2.24 would not be rigid. To emphasize the point

11. I will ride this hobby-horse—the combination of utility and aesthetic appeal via applied science—again in this book, and more than once.

that beams are subjected to bending moments in the roof designs of figure 2.24, I have placed pins at those joints where they *are* permitted—the structures are rigid with the pins placed as shown.

Earlier, I mentioned in passing the extra complexity that is part and parcel of three-dimensional truss structures (fig. 2.9) compared with the two-dimensional trusses that I have analyzed in this chapter. Not only is

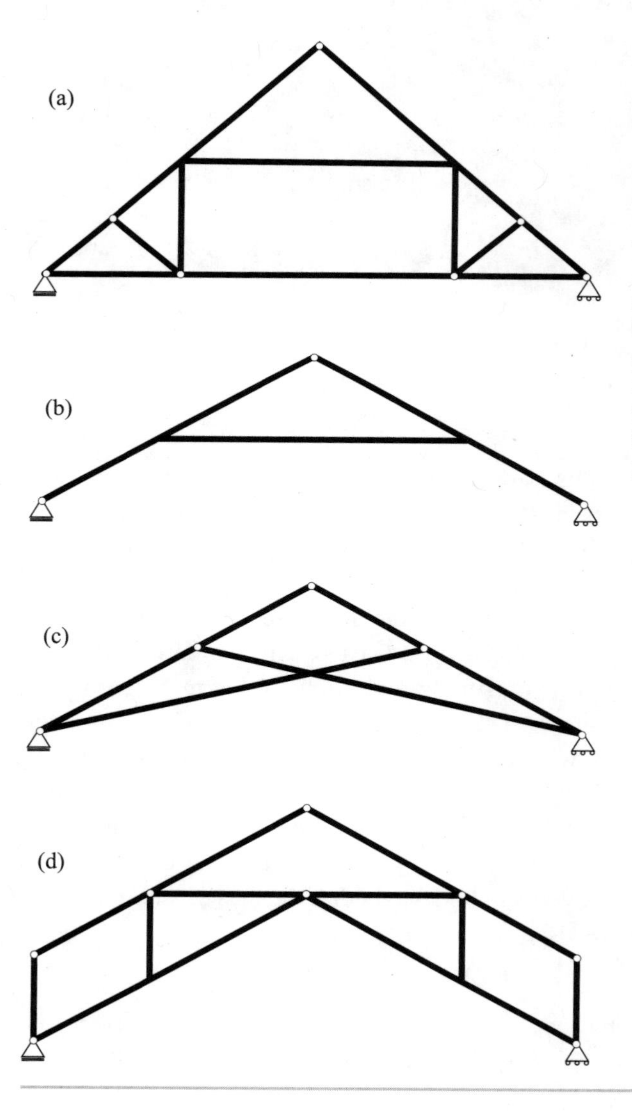

FIGURE 2.24. Pitched roof trusses, with beams: (a) queen post, (b) collar beam, (c) scissors, and (d) hammer beam. If the joints marked by white circles were replaced by pins, the structures would still be rigid (the pin placements are not unique). However, they would not be rigid if the remaining joints became pins.

FIGURE 2.25. Westminster Hall, London. Image courtesy of the House of Commons Information Office, London.

the analysis more complex, because each joint gives rise to three equations instead of two, but there must be extra members added to 3-D structures to ensure stability. The 2-D roof trusses of figure 2.22 are stable in the plane of the page, but to form a roof they must be placed in a row along the third dimension and then stabilized. Extra bracing is needed to make this 3-D roof structure stable. I will not analyze 3-D trusses in any more detail here, since no new physical principles are brought to light that might justify the additional mathematical complexity.[12] Instead, I end this chapter by referring you to figure 2.25, which illustrates the point that I touched upon a few pages ago. Art and science, aesthetics and engineering come together in all the big structures of this book; here is another example in the form of a hammer beam roof. The two examples given here (figs. 2.23 and 2.25) are centuries old, which shows us in both cases that the design and the timber are solid. Also, both are visually appealing.

12. Not quite true. Torsion adds extra complexity to 3-D structural analysis, but in the context of trusses this twisting force does not exist: only axial forces act along the length of each member.

SUMMARY: This chapter—the most mathematical of the book—has given you a grounding in simple truss theory. This theory ignores internal forces such as bending or shear and considers only the external tensile or compressive forces that are brought to bear upon the members of a truss structure. Some constructions of truss members are statically indeterminate, meaning that we cannot ignore internal forces that act upon the members. The ease of truss calculations (compared with, say, beam theory) gives us insight into the forces that act upon large structures. We have also applied truss theory to simple bridges and roofs.

CHAPTER 3

Towers of Strength

Why Build Tall?

"At long last," you say to yourself, "he is finally going to tell me about big buildings and structures." Yes, indeed, but no apologies for spending the first two chapters telling you about building materials and truss theory because these have set us up well for the investigations that will follow. I can now begin to describe to you some of the interesting science and engineering, some of the astonishing facts and figures, concerning the first of our large structures: tall buildings and towers.

There are practical reasons why individuals and organizations wish to build tall structures. Smokestacks and cooling towers are made high so that the smoke or steam emanating from them is carried far away and dispersed by the higher-speed winds that are found aloft. Lighthouses and communications masts are built high to overcome the curvature of the earth; visible light, radio waves, and microwaves travel in straight lines, more or less, and so higher antennas can communicate farther.[1] Two man-made high-rise structures, old and new, are shown in figure 3.1. Another reason for building upwards is real estate cost. In some cities the price of land is so high that it becomes worthwhile to build higher and higher, despite the spiraling price tag for doing so.[2] The main reason, however, that people, organizations, and

1. Atmospheric refraction can cause visible light, radio waves, and microwaves (all of them examples of electromagnetic radiation, differing only in frequency) to deviate slightly from straight lines and instead travel along curved paths. This refraction is due mostly to variations in atmospheric density with altitude. A mast that is H feet above sea level can communicate a distance of about $1.35\sqrt{H}$ miles with a transmitter or receiver at sea level. So, two 100-foot masts can communicate a distance of about 27 miles over the ocean. (On land, the presence of mountains makes the determination of communication range more difficult.)

2. Edinburgh, Scotland, developed high-rise buildings (up to 14 stories) in the

nations build tall buildings and towers is prestige. Large corporations vie for the tallest skyscraper headquarters; cities and nations proudly declare a tower to be their icon. Thus, in New York City we have the Empire State Building and the Chrysler Building; until the tragedy of 9/11 we had the giant World Trade Center Twin Towers as symbols of world trade (the replacement Freedom Tower will be even taller). Chicago boasts the Sears Tower; Paris, the Eiffel Tower; Kuala Lumpur, the Petronas Towers. I will have something to say about all of these structures and others like them in this chapter.

Flying Buttresses

The most common example of a tall tower in classical history is the church steeple.[3] Built high to make a statement—to be imposing—the steeple also inspired awe among the faithful who flocked to worship. The church interior reinforced this sense of awe by enclosing a large and (importantly) a high space, particularly if it was a cathedral or an abbey—a center of ecclesiastical power. The cruciform layout of traditional church architecture includes a *nave* and a *transept* that cuts across it, forming a cross. The nave may have *aisles* along each side. In larger churches these aisles may not carry up to the same height as the nave; they may be shorter, in which case *clerestory* windows in the nave walls above the aisles let in more light, adding to the striking effect of the nave height.

Traditionally, cathedral steeples and naves were usually built of stone. We have seen that stone is weak in tension, and so the cathedral design must be such as to avoid tensile stresses upon the walls. These stresses naturally arise, however, within tall (and therefore heavy) structures, especially when the architect wants to create an uninterrupted high space within. We can understand why this is so by turning to a simple truss model. In figure 3.2a a basic truss (a tall church roof or steeple) is subjected to a downward load, W, representing the roof weight. This truss has been analyzed in note 3 of the appendix, recall.[4] We found that the load compresses

---

seventeenth century because of the citizens' desire to live within the city walls and the consequential high cost of land.

3. During the Middle Ages, Lincoln Cathedral in England was the tallest manmade structure for over 200 years. Currently the tallest church is Münster Cathedral in Germany (530 feet).

4. If you are following the math, you will see that the analysis of appendix note 3 applies to the case shown in fig 3.2 if we set the triangle angles $a = b = \theta$. This analysis produces the forces shown in fig. 3.2.

FIGURE 3.1. Towers, old and new. (a) The square tower of Corunna lighthouse, on the north coast of Spain. I am grateful to Waldemar J. Poerner for this photo. (b) The KL Tower in Kuala Lumpur, Malaysia. Thanks to Peter Lee for this image.

both diagonal members and induces a tensile stress in the horizontal member. Now suppose that we remove the horizontal beam, to create a larger uninterrupted space between the congregation below and the high rafters above. The horizontal forces that were held in check by the beam will now act to push the walls apart, perhaps catastrophically. So, we need to counter these forces in some other way. The traditional method is to add *buttresses* to the outer walls. Buttressed walls are shown in figure 3.3, and schematically in figure 3.2b, where the forces that act upon the wall are shown.

To be effective, the buttress on the right in figure 3.2b must be wide enough so that load is directed through its base and not through its side. (I derive the requirement mathematically in note 4 of the appendix.) Sometimes the aisles of churches, being less high than the nave, can act as buttresses; this works better aesthetically because it hides, or rather disguises, the somewhat ugly buttresses.

A more sophisticated method of buttressing—more elegant and effective, though more difficult to construct—is via the *flying buttress*, also shown in figure 3.2b. Real flying buttresses appeared in European churches and cathedrals during the Middle Ages; one important example is shown in figure 3.4.[5] (Some buttresses were required *inside* the church; a famous example is Wells Cathedral in southwestern England, shown in figure 3.5.) Once again, the problem of countering the horizontal forces using stone, which is strong only under compression, is addressed, but here the solution requires less horizontal space (this fact is demonstrated in note 4 of the appendix) and is more architecturally appealing. Historically, the tall flying buttress columns were ornately decorated to enhance aesthetic appeal and add to the cathedral's splendor.

Truss Tower

So much for medieval steeples; how do tall towers work in general? We can gain some insight into the stresses that act upon a tall tower by turning once again to truss theory. In figure 3.6a we have a rather complicated truss

5. Chartres Cathedral also sports splendid flying buttresses. The French preferred flying buttresses, whereas the English tended to adopt a different solution—the hammer beams that we saw in the last chapter. Historical architecture tended to be *additive* in the sense that, if a building was seen to be developing cracks or alarming bulges, it would be shored up with buttresses of one sort or another. More recent architecture is *deductive*, meaning that we can use our engineering knowledge to anticipate problems and solve them on the drawing board.

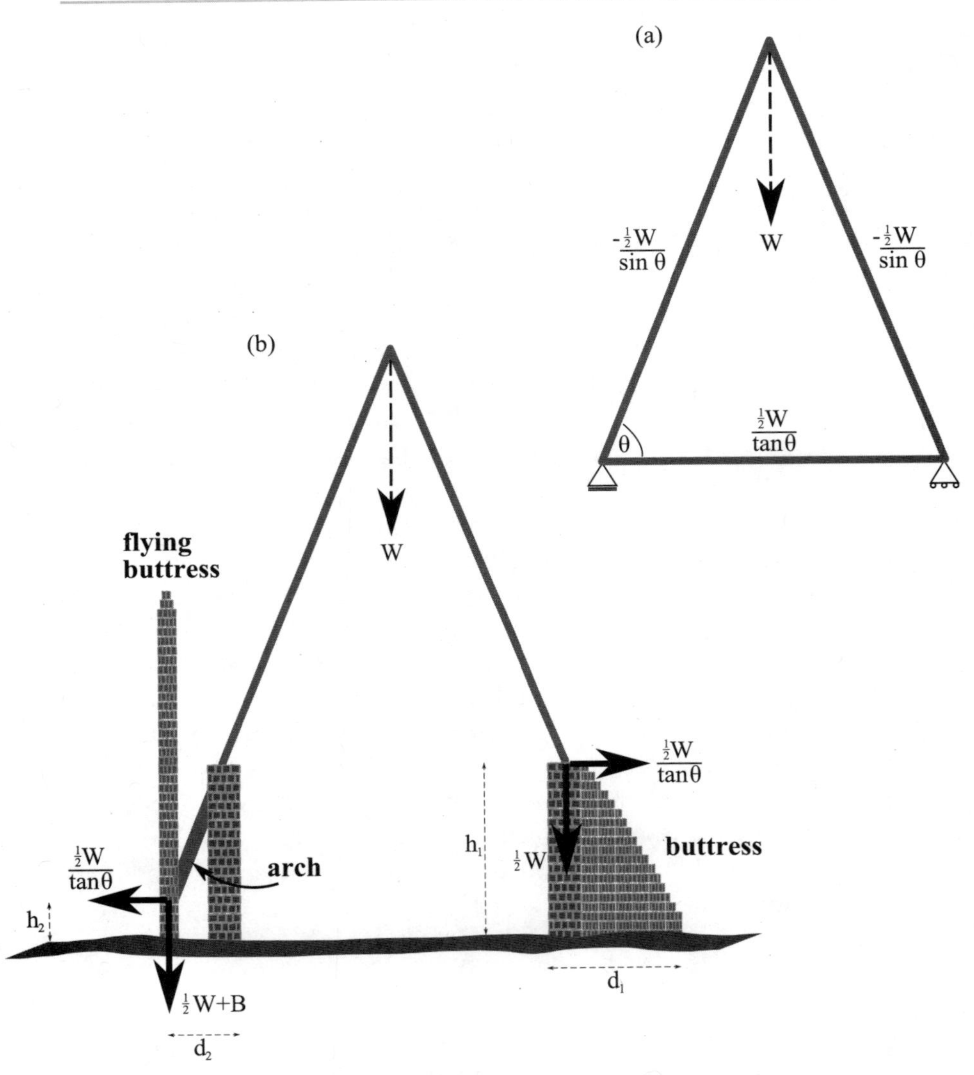

FIGURE 3.2. Tie beam versus buttress. (a) A triangular truss roof. The weight *W* compresses the diagonal members, but stretches the horizontal member. (b) If the horizontal member is removed, then the force pushing outwards has to be countered in a different manner. On the right we see a buttress attached to the wall, and on the left a flying buttress, separated from the wall by an angled arch.

tower—it may be a TV antenna, a communication mast, or the metal frame of a high-rise apartment block. I have assumed for this truss tower that the weight, w, is divided equally throughout the structure as shown (a vertical load of $w/20$ is applied at 20 pin locations). Similarly, I have assumed a wind loading of p that is uniformly distributed up one side of the tower (a horizontal load of $p/10$ at 10 locations). In figure 3.6b I have calculated the tensile and compressive stresses that act upon the outside truss members. (See how much I labor for you?) The interior stresses have also been determined for this structure, but the figure is already cluttered enough

FIGURE 3.3. High-rises, old and new. Buttresses supporting the steeple walls of St. Patrick's Cathedral, New York, are clearly visible. Thanks to John R. Plate for this image.

FIGURE 3.4. To call these flying buttresses "medieval scaffolding" would be structurally correct but architecturally boorish. This is Notre Dame in Paris, and the buttresses add aesthetic appeal along with structural support. They permit extra windows in the walls, which otherwise would be too weak to permit windows. Thanks to Don Cooper for this image.

FIGURE 3.5. The interior of Wells Cathedral in southwestern England. The scissor buttressing was added in the fourteenth century, long after the building was completed, to fix a problem—the walls were caving inward. The fix worked, and the buttresses have been integrated magnificently into the cathedral design.

without them, so they have been omitted for clarity. Recall that compressive forces are negative and tensile forces, positive. So, the leeward side of the tower (downwind) feels a stronger compressive force than it would if there were no wind, whereas the windward side feels a reduced compressive force. In the absence of wind the forces are symmetrical. All this is to be expected. The main point to note is that the effects of the wind loading are felt particularly strongly in the *lower* part of the tower.

Though informative, this model is perhaps a little too cluttered to be easily digested, and so I will simplify further by modeling our tower as the simpler truss structure shown in figure 3.6c. This model lacks the detail of the other tower (for example, we do not see how the stresses vary with height above the ground), but it still gives us an overall impression of the stresses that are felt by the truss members due to the wind and weight loading shown.[6] In particular, we can see quite easily how the stresses vary with tower height, via the angle θ. For tall towers, θ approaches 90°; as a consequence, the wind loading stresses become very large indeed. It is quite realistic for a TV mast, for example, to be 50 or 60 times as high as it is wide; in such a tower $\theta = 89°$. So the wind loading is magnified by a factor of $\tan(89°) = 57.3$ in the leeward vertical member. (In practice, many slender TV masts have cables supporting them so that wind loading is less of a problem.) Wind loading is indeed a significant concern for tower designers, as we will see, and my simple truss tower model shows us why: the magnification of wind force is much like the effect of a lever. Imagine pushing with a horizontal force p applied to the top of a crowbar that is wedged into the ground. The lever effect would magnify the force you apply—and the longer the crowbar, the greater the lever effect.

Wind force increases as the square of wind speed, and so the maximum wind loading that a tall structure can take will correspond to a critical wind speed. You may have seen reports of extreme weather conditions in which high-speed wind gusts have been held responsible for structural damage. The type of damage depends upon the structure. A TV mast made of light steel members may buckle on the lower leeward side, as we have seen. A tall and very thin brick or stone structure, such as an old chimney stack, may fail on the windward side where the normally compressive stresses may become tensile in a strong gust of wind (see fig. 3.6b, or the diagonal member of fig. 3.6c).

6. This truss is simple enough that you may be tempted to check the calculation for yourself if you are feeling virtuous.

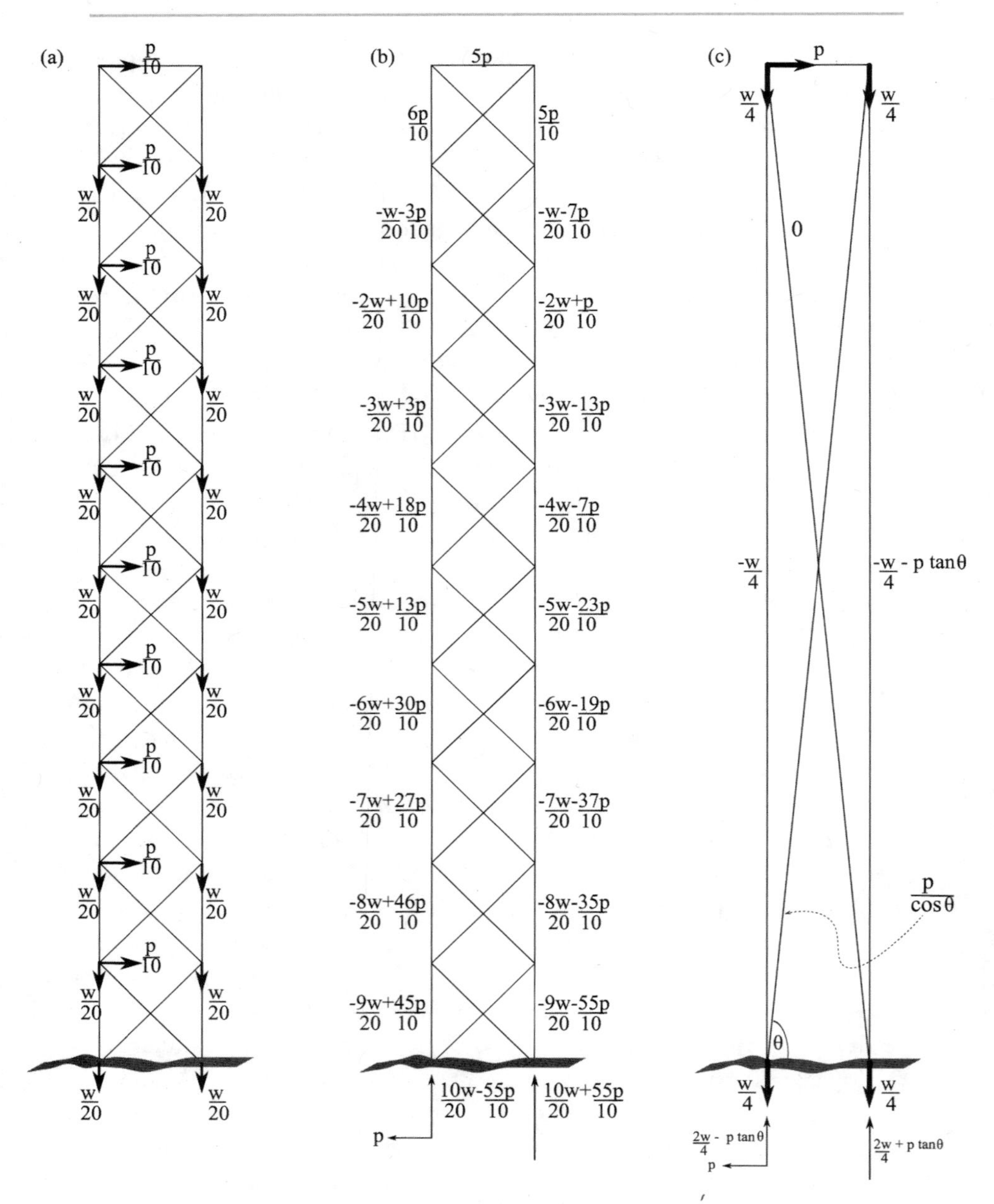

FIGURE 3.6. The forces that act on a tall tower. (a) The wind and gravity loading on a truss tower. (b) The tensile and compressive stresses that act on the truss members. (c) A simplified truss tower, with forces and stresses shown.

Apart from simple triangular trusses, it is possible to make strong tall towers from long, straight steel beams. You may have seen the cooling towers at power stations, strangely and elegantly curved—narrower at the waist than at the top or bottom. Until I learned about *hyperboloid* towers, I had always assumed that this strange shape was something to do with increasing the efficiency with which gases escaped up the tower. In fact, the hyperboloid is a mathematical shape that can be constructed from straight beams, surprisingly enough, and so it is economical to build. The explanation is provided in figure 3.7a. A Russian engineer by the name of Vladimir Shukhov gave us the first hyperboloid tower at the end of the nineteenth century (see fig. 3.7b), and he made quite a living by construct-

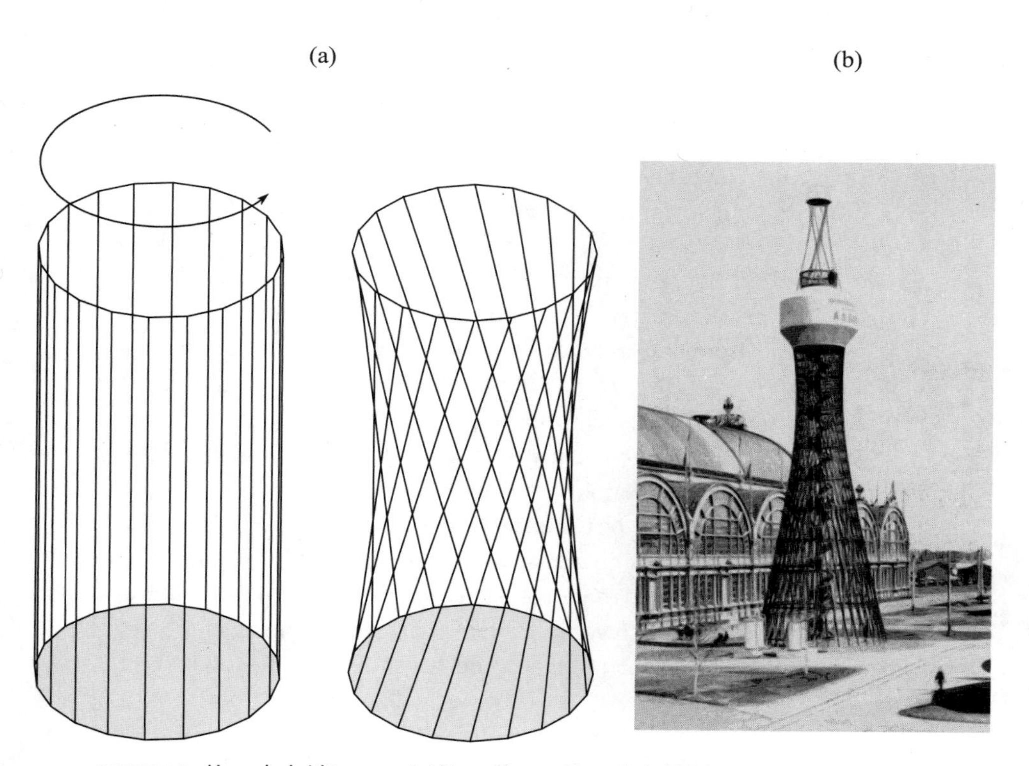

FIGURE 3.7. Hyperboloid towers. (a) To see how a hyperboloid is constructed from straight beams, consider the structure on the left: two disks are connected by metal wires or by string. The top disk is then twisted, resulting in the shape shown on the right—a hyperboloid. You can see how steel reinforcing bars might be made into this shape to form the skeleton of a reinforced concrete cooling tower. (b) The first hyperboloid tower, built in Russia in 1896. Image from Wikipedia.

ing many such towers.[7] The mathematics of hyperboloids had been appreciated much earlier; in fact, the seventeenth-century London architect and mathematician Sir Christopher Wren made a significant contribution to this understanding, though we had to wait for a couple of centuries to see the practical structural consequences of hyperboloid mathematics. A number of prestigious buildings and tall structures around the world are based upon the hyperboloid—for example, Brasilia Cathedral in Brazil and the planned 2,000-foot tower in Guangzhou, China, which will have a sightseeing platform and is due to be opened in time for the Asian Games in 2010. The McDonnell Planetarium in St. Louis, Missouri, has a striking hyperboloid roof. The now-demolished Gettysburg National Tower, in Pennsylvania, was a hyperboloid observation tower, as is the 350-foot Kobe Port Tower in Japan.

Eiffel Tower

If you don't recognize the subject of figure 3.8, then you must have spent your life with your head in a bucket. The Eiffel Tower is one of the most visited tourist destinations in the world: over 230 million people had visited it as of 2006, according to the Eiffel Tower official website. This tower is an icon, not just of Paris, but also of the new industrial age in France.[8] It was intended to be exactly that: a symbol of modern France, embracing the new age of iron. Built for the 1889 Expo, celebrating the centenary of the French Revolution, the Eiffel Tower is exactly contemporary with the Firth of Forth Rail Bridge near Edinburgh, Scotland. Both edifices are built of iron scaffolding that is held together by rivets—2½ million of them, in the case of the tower. Apart from its iconic status as a symbol of Paris that is recognized anywhere in the world, the Eiffel Tower has at various times served more typical tower functions. In times of war it has acted as both a radio tower and a lookout post. In times of both peace and war, it has served as a billboard and flagstaff. In 1940 following the

7. Shukhov also designed and constructed the first tensile steel shell roof. Tensile roof shells are now commonplace, covering large spaces such as sports arenas. They require a high-tech product (high-tensile-strength steel) and therefore became possible only during the last 100 years or so.

8. Interestingly, there was some opposition to the building of this tower when Gustave Eiffel first proposed it in the 1880s. Many Parisians felt that it would be an ugly high-rise eyesore. Eiffel persuaded them that it would be artistic as well as impressive.

FIGURE 3.8. *Top*: Despite its 7,300 tons of iron, the Eiffel Tower is lighter than air, in one sense. See text for details. *Bottom*: Iron trusses, rivets, and air—the Eiffel Tower from underneath. I thank Don Cooper for these images.

German occupation, the tower's elevators were disabled, and the victorious Germans had to ascend the 1,665 steps[9] to hoist their swastika flag. Four years later, the elevators were back in working order within days of the German evacuation.[10]

The tower is about 1,000 feet high—a little higher or lower depending upon whether you include the antenna at the top. The four legs form a square that is about 350 feet on a side. The total weight of the Eiffel Tower approaches 10,000 tons; of this, 7,300 tons is in the form of cast iron, while the rest is stone foundations plus the large (and heavy) elevator system. Wind can cause the tower to lean but only a little—less than 3 inches. Much larger, at 7 inches, is the lean caused by differential thermal expansion. The iron on the sunny side of the tower expands on a hot day more than the iron on the shady side and pushes the tower over a little bit.[11]

Because it is made of iron, the Eiffel Tower needs to be painted regularly to prevent rusting (the Forth Rail Bridge also requires near-continuous painting)—50 tons of paint every 5 years, applied by brush. The iron scaffolding (a series of trusses, as you can see from fig. 3.8) must be extremely tedious to paint, but such is the price of ironwork. Looking at figure 3.8 you will be forgiven for thinking that the Eiffel Tower is monumentally

9. Approximately. Accounts differ; some say 1,792 steps, and some say other figures. While I always strive for accuracy, I have no plans to settle this matter by ascending the steps and counting them.

10. Hitler ordered that Paris be left a ruin by the retreating German army, and the order included the destruction of the Eiffel Tower. The German governor of Paris, von Choltitz, refused to carry out this order.

11. It is not difficult to confirm that this is a reasonable claim. We can model the top part of the tower—the section above the second platform—as a giant inverted V, made of iron. The height of the V is about 675 feet and the base is about 80 feet. So, the angle to the horizontal of each leg of the inverted V is about 87°. This inverted-V structure is close enough to the Eiffel Tower for our rough, back-of-the-envelope calculation. The coefficient of thermal expansion for iron is 0.0000111 C^{-1}, meaning that 1 foot of iron will expand by 0.0000111 feet for each degree centigrade increase in temperature. Thus, one leg of the V will increase in length by about 0.09 inches more than the other leg for a 1°C difference in temperature between the two legs. A 4.5°C difference leads to a length difference of 0.4 inches.

From the geometry, it is not too difficult to show (I will leave this as an exercise for the interested student) that the top of the V will shift toward the cooler side by an amount 0.4/cos(87°), or just over 7 inches. A temperature difference of 4.5°C (say 8°F) between the two sides of the V seems plausible; thus, the stated swing of the V, and of the Eiffel Tower, due to differential thermal expansion also is believable.

heavy. Well, literally speaking, I suppose it is, but in a sense the tower is lighter than air. One claim made on the official website is quite intriguing and is worth pursuing: if we could make a giant cylindrical box that contained just the Eiffel Tower, the air in the box would weigh more than the tower. This claim is true,[12] and consequently, the average pressure exerted by the tower upon the foundations is less than air pressure. (The pressure beneath the four legs of the tower is higher than the average; even so, it is only about 64 psi—a compressive stress that the stone foundations can easily tolerate.) The stress at any point in the tower framework is no more than 11 psi.

From the structural engineer's point of view it is interesting to note that during the renovations of the Eiffel Tower in 1980–85, engineers strengthened the structure and removed 1,343 tons of what is described as "unnecessary material." Given what we learned in chapter 2 about truss theory, I assume that this unnecessary material consisted of zero-force truss members. Using modern measurement techniques and computer analysis tools, engineers nowadays can quantify the stresses in complicated structures such as the Eiffel Tower quite accurately, whereas the original designers had much less precise knowledge of the forces that act on given members, and so they would have felt obliged to build in a large safety factor, in the form of redundant zero-force members.

When asked about the shape of the tower, Gustave Eiffel said that this was determined by wind loading calculations. Perhaps so: we have seen how wind loading can upset the neat and symmetrical loading that results from gravity alone, and how it can be very significant for tall structures. It may, indeed, determine the material—strong in compression—that we use to build our structures. Wind loading may place some members in tension, and that is not good if the members are weak in tension. In fact, we can show from a simple truss model of the Eiffel Tower that the shape may have been determined by a desire to minimize the tension that arises in some members even without wind loading. Again, the hard work of chapter 2 in coming to grips with truss theory pays off; we can apply the theory in

12. Again, we can perform a back-of-the-envelope calculation to show that this surprising claim is reasonable. The hypothetical cylinder would have a base radius of about 290 feet, a height of about 1,000 feet, and so a volume of 264 million cubic feet. The density of air is approximately 1.2 ounces per cubic foot, so the weight of air in the cylinder is about 8,660 (long) tons—more than the weight of iron that constitutes the tower. If we were to put the tower into an oblong box, the weight of air would be a mere 5,600 tons.

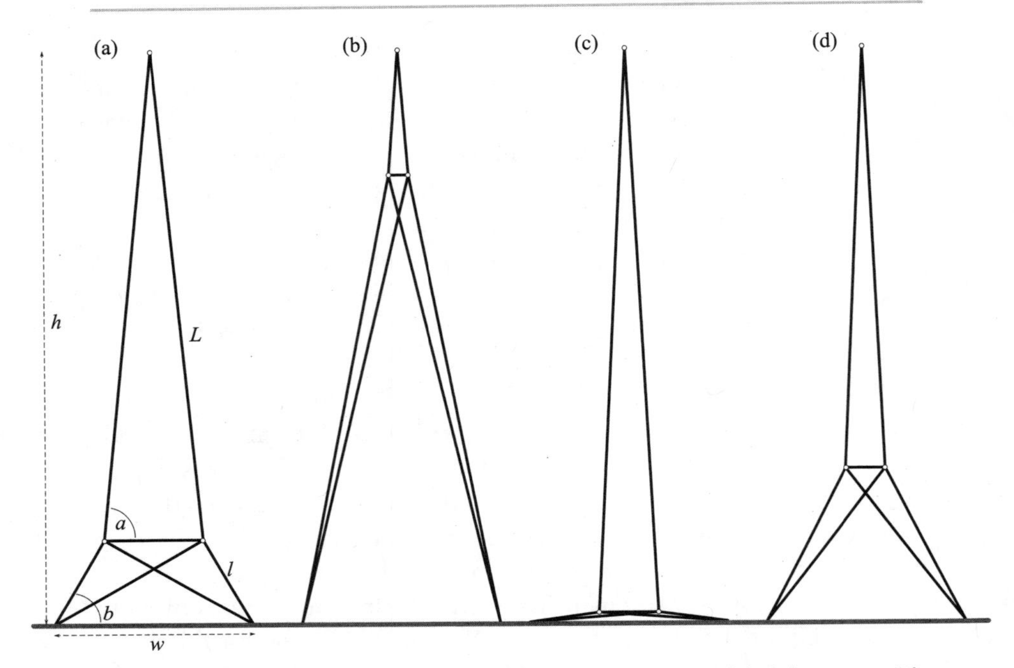

FIGURE 3.9. Truss models of the Eiffel Tower. (a) A simple truss model of the tower with the same ratio of width, *w*, to height, *h*, as the actual structure. We fix angle *a* and let angle *b* vary. (b) One extreme case, for very large *b*. (c) The other extreme, for very small *b*. (d) The case with minimum tension stress in the diagonal members.

simple ways to different structures to gain insight into the forces that act upon those structures.

From Eiffel's blueprints we can measure the separation of the tower legs, the tower height, and the angle of the legs (for this 2-D projection). The ratio of tower height to width is $h/w \approx 3$, and the leg angle is about 65° from the horizontal. Figure 3.9 shows my simple truss model of the tower. It has the same height-to-width ratio as the Eiffel Tower, and I will assume that the height, h, and base width, w, are fixed constants. The truss tower has seven cast iron members and five pins, including two on the ground; note that the diagonal members are connected to the ground but not to each other. The weight, W, of the tower is assumed to be evenly distributed, so that each pin is subjected to a load of $\frac{1}{5}W$. The truss is statically determinate, and I have calculated the stresses that each member feels due to the load.

For the calculation I assumed that the angle a shown in figure 3.9 is fixed at $a = 87°$ (so that the top of the truss tower closely resembles that of the Eiffel Tower, fig. 3.8). Angle b is allowed to vary between the two extreme cases shown. The member lengths L and l are determined by w, h, a, and b. It turns out that the diagonal members feel a tensile stress, whereas the other three members feel a compressive stress, whatever the value of angle b. We know that cast iron is very strong under compression, so we are not concerned about the compressive loadings. We would like to minimize the tensile loading in the diagonal members, however, because cast iron is weak under tensile stress, as we saw in chapter 1. The value of angle b that minimizes the tensile stresses is $b = 63°$, and the truss tower with this optimum angle is shown in figure 3.9d. Note how similar the shape is to that of the Eiffel Tower. The optimum b is very close to the measured leg angle of the real tower, 65°. So, from this simple model we conclude that the tower shape may have been chosen to minimize the tensile stresses acting upon the members.

Of course, my Eiffel Tower truss model is very simple (it is a 2-D model, for example), and perhaps it is too simple to make definitive, quantitative conclusions—but the results are very suggestive. You see how we can gain confidence in the application of simple truss models of much more complicated real structures; these models provide only approximate results, but they give us a good sense of the stresses that act within the real structure.

The Sky Is the Limit

Skyscrapers were born in Chicago in the second half of the nineteenth century, and with the Sears Tower, Chicago still boasts one of the world's tallest buildings. In the early twentieth century New York came to symbolize skyscrapers throughout the world. A well-known architect from this period, Louis Sullivan, wrote of the "glory and pride" that these buildings bestowed upon the builders: "The man who designs in this spirit . . . must be no coward, no denier, no bookworm, no dilettante. . . . He must realize at once and with the grasp of inspiration that the problem of the tall office building is one of the most stupendous, one of the most magnificent opportunities . . . ever offered to the proud spirit of man." Heady stuff—quite literally, I suppose, when you stand atop one of these creations.

The type of spirit that urges people to build skyscrapers has spread around the world, as we will soon see. The buildings just keep getting

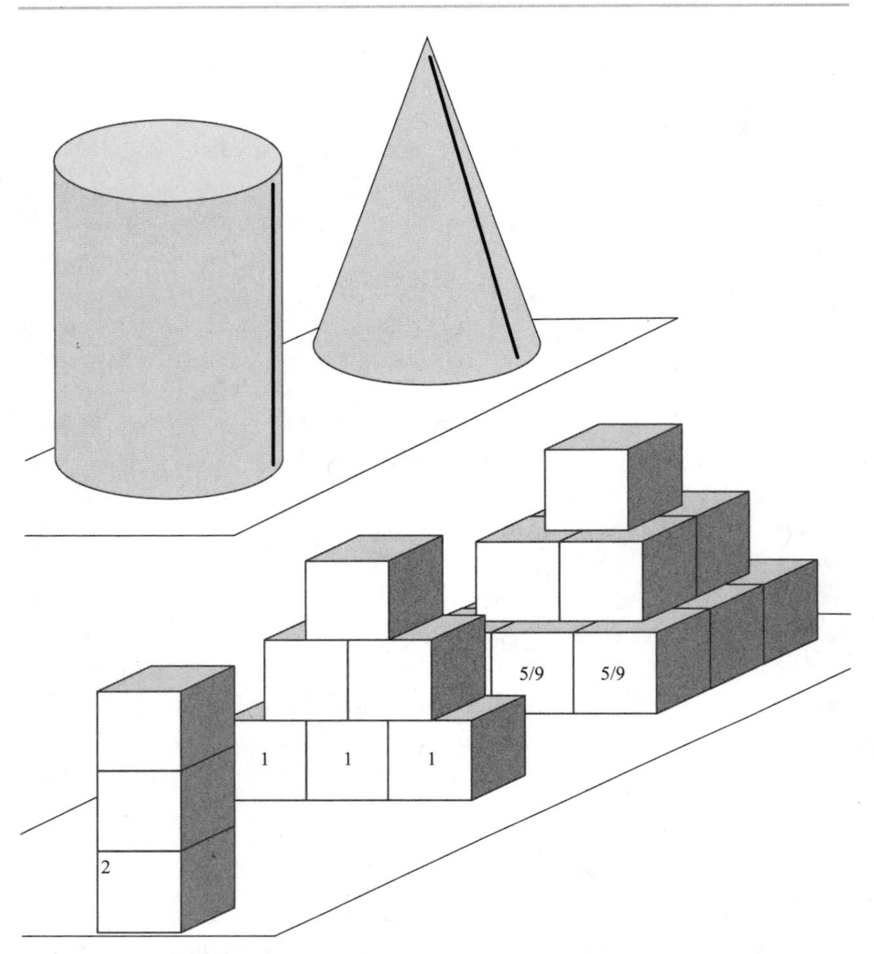

FIGURE 3.10. Building shape and compressive stress. The cylinder and cone have the same height and base area, but the cone volume is one third that of the cylinder, and so the pressure at the base is correspondingly reduced. Similarly, the three apartment blocks shown have the same height, but the compressive stress on the lowest level is less for the pyramid than for the tower. The numbers show the average weight (in number of blocks) borne by each ground-floor block.

bigger. The record in November 2007 stood at 1,700 feet (plus or minus, depending upon whether or not you include the antennas that inevitably adorn skyscrapers). At that time there was some disagreement as to which building held the record: it depends upon definitions, upon which superstructures are considered to be an integral part of the building and which are not. Since then, the record has been broken—shattered, in fact—by the Burj Dubai building in the Arabian Gulf, which is due for completion in September 2009 and will soar over half a mile into the air. As of January 2009 the Burj Dubai had reached its final height of 2,684 feet, though the building was not yet completed.

Just how tall can a building be? Is there a technological limit? We saw in chapter 1 that the compressive strength of building materials imposed a limitation, though not a very strict one. Concrete buildings, we saw in table 1.1, could be 4,800 feet high before the lowest blocks crumbled under the heavy load above them. Steel buildings could be three times higher. This limitation—compressive strength—is not the last word, however, because the calculated maximum height assumes that the building has a uniform cross section, as sketched in figure 3.10. If the building tapers, as for a pyramid or cone, then the load on the lowest blocks is spread wider and the compressive stress is reduced, for a given building height, as shown in the figure. So, the maximum height to which buildings can be built, in practice, will not be limited solely by the compressive strengths of building materials.

The earliest skyscrapers appeared in the late 1800s. The outer (curtain) walls of these buildings were made of stone, and they carried the load. That is to say, the stone blocks that made up the fifteenth story carried the weight of everything above the fifteenth story; the blocks of the second story carried the weight of everything above the second story. The next generation of skyscrapers included the famous Art Deco exteriors of the Empire State Building and the Chrysler Building. These structures have a steel core that bears the heavy gravitational load; the elegant walls support only their own weight. For this reason, the walls could be opened up to improve the view. Third-generation skyscrapers (many of them dating from the 1960s) have a more complex skeleton consisting of several vertical sections. This complexity permits greater external variations—for example, variation of shape with height. Such variability helps to reduce wind loading, which is one of the main problems associated with tall buildings, as we have seen.

The horizontal load imparted by wind increases as the square of wind speed, and wind speed increases with height above the surface. It is not difficult to imagine that a building that stands so much taller than those around it will experience serious horizontal forces due to the wind. Detailed aerodynamic analysis shows that the problem is due to "vortex shedding," the detachment of swirls of wind from the leeward side of a building.[13] This phenomenon is more marked in buildings of a uniform cross section—one that does not vary with height. Therefore, varying the building shape with height reduces the wind loading. Such variations are part of the process of making the building shape more aerodynamic. Just as airplane designers choose the cross-sectional shape of airplane wings very carefully to reduce drag, so designers of the latest generation of skyscrapers are sculpting the shapes to minimize wind loading. In these high-rises the sharp corners seen in traditional buildings are smoothed, so that the square cross section becomes an octagon, for example, or an ellipse. Rotors can be added to buildings at strategic places to control the air flow. Sometimes holes are placed in a tall building, especially near the top, to let the wind through. Such designs are feasible with the latest generation of strong building materials and building designs. A tall building with a hole in it is visually striking and can be architecturally dramatic, but the main reason for the hole is to reduce wind loading.

Another challenge that acts to limit the attainable height of skyscrapers is elevator logistics. Elevators are essential for skyscrapers, of course, and elevator technology has evolved in step with skyscrapers to solve the problems of entering and leaving the upper floors in a safe and rapid manner. Modern elevators are much faster than the older designs and can speed passengers up a 100 stories in less than a minute at speeds exceeding 50 feet per second; the latest elevators are double-deckers. Increased elevator speed and capacity help to reduce the number of elevator shafts required to access the upper floors. A large number of elevator shafts greatly reduces the useful floor space and results in a compromise between skyscraper height and floor space, defeating one of the reasons for building high in the first place. Thus, much thought has gone into elevator logistics of tall buildings. For example, both the Twin Towers of the World Trade Center had three-tier elevator systems; passengers would change elevators part way up, much as they change trains in a metro. From the elevator point of view, the WTC Towers were each three buildings stacked vertically.

A skyscraper's skeleton—its steel or concrete core[14]—sits on a large

13. When vortex shedding occurs at the trailing edge of an airfoil, the aerodynamic drag force that acts upon the airfoil increases. Indeed, to a fluid dynamicist, vortex shedding and drag are intimately related.

14. Tall buildings that are designed to hold a high "live load," such as apartment blocks, are usually made of concrete because it is stiffer than steel and provides better soundproofing. Commercial buildings have a much lower live load; these are usually constructed of steel.

concrete foundation pad, and this substructure may in turn rest upon giant piers that penetrate down all the way to bedrock.[15] So, the skyscraper base is firmly anchored, but the top is not. A skyscraper in a strong wind floats around like kelp in an ocean current. This image is hyperbole, of course—the building does not move so much or so freely as a string of seaweed anchored on the sea bed—but it makes a point. Buildings, and especially very tall buildings, do sway in the wind. We have seen how the Eiffel Tower moves several inches; modern skyscrapers are much taller, and with a bigger profile, they can sway several feet. Such swaying is not necessarily unsafe, but it can be uncomfortable (and perhaps unsettling) for people on the upper floors. So, stiffening is required. *Shear walls* add rigidity to the shell and reduce the swaying, but a very tall building needs more than shear walls to attenuate the swaying motion. The most recent generations of tall buildings have increasingly sophisticated techniques for damping the sway motion due to variable wind loading. These techniques are many and varied, ranging from passive mechanical to active computer-controlled feedback systems that deal with earthquake vibrations as well as wind gusts.

The oldest and most obvious method of reducing building sway is to stiffen the structure, by, for example, adding extra members to a steel truss core. Another method is to place a water tower at the top of the building (this may be convenient anyway, for water distribution and fire suppression). The water will slosh around inside the tank as the building sways; with a big tank the sloshing water can noticeably reduce swaying. A more sophisticated version is the tuned mass damper: computer-controlled hydraulics push a large concrete block weighing hundreds of tons back and forth along a top floor, in such a way as to diminish the building sway. The feedback system measures wind loading, calculates the best position for the concrete weight, and then moves the weight. Some buildings have a giant pendulum suspended from an upper floor; the pendulum bob, weighing hundreds of tons, moves naturally in such a way as to lower the sway frequency. It seems that occupants are more sensitive to higher oscillation frequencies in buildings than they are to lower oscillation frequencies, and so lowering the frequency is a perceived benefit.

The problems of building motion will become more acute as buildings get taller because wind loadings will increase and because taller buildings

15. Skyscrapers in Japan are much more expensive to build than skyscrapers in Chicago because bedrock is much deeper in Japan.

sway more than shorter ones—but also because the new strong materials that we use to construct such buildings are lighter and therefore more susceptible to being moved around. One very high-tech approach that may be used in tall buildings of the future, for earthquake protection as well as wind load damping, is based upon *magnetorheological* fluids. These fluids (mercifully abbreviated to MR) have the strange property that they change from a liquid state to a nearly solid state when subjected to a strong magnetic field. The idea is that MR fluids could be placed in large pistons attached to the spine of a skyscraper skeleton and then could stiffen the building frame if required. This stiffening could be applied very quickly, by switching on a magnetic field, and it could be applied selectively at different parts of a building, depending upon the nature of a sudden earthquake jolt. All objects, including tall buildings, have one or more resonant frequencies that amplify the swaying or vibrating when a force is applied—a force that just happens to hit that resonant frequency. (Resonance and other oscillation phenomena are briefly recapped in note 5 of the appendix.) The mass damper, MR fluid, and other damping systems, some of which are still on the drawing board, can reduce the magnitude and duration of these resonant oscillations, as well as alleviate the effects of transient loads such as are produced by earthquakes.

In summary, we now have the technical know-how to build tall structures, the height of which is not limited by material strength or by structural design constraints. Conventional elevator systems limit height (to about 80 stories, it seems), but new elevator designs and novel elevator dispositions (such as elevator stacks and exterior elevators) overcome this limit. Wind loading is an increasing problem as building height increases, but damper systems (along with new, strong, and very stiff construction materials) will control the problem. Not all tall buildings today have damper systems because not all tall buildings need them, but damper systems will be needed for the supersized structures of the future. So what will be the ultimate limiting factor? The great American architect Frank Lloyd Wright thought that there would be no limits. He overcame his initial dislike of skyscrapers and designed a mile-high pipe dream to be called the Illinois Mile High Tower. This building never got beyond the design stage: it never could in 1956, when Wright drew up his design. He felt that all the problems we have mentioned here, except possibly for the elevator problem, could be overcome. Other issues that arise with very tall buildings (such as differential thermal expansion, water pressure, and fire safety) are also not fundamental limitations.

The ultimate brake on skyscraper height will be imposed by cost. The cost of these buildings is astronomical; per floor it is greater than for other buildings, and the number of floors is greater. Cost may be driven by real estate prices, in which case increasing the base area of a skyscraper may not be an attractive option. If the skyscraper is to contain high-end, high-rental, high-prestige, and just plain high dwellings, the renters may be put off if they can see nothing below them but clouds. In many parts of the world, a mile-high apartment building may be unrealistic for this reason alone. The high cost of getting large and heavy structural components up to the top of a tall building that is under construction will also deter, eventually.[16] If buildings have been made tall in the past for reasons of kudos, prestige, or even ego, the financial cost of these human foibles will one day prove too high. How high the skyscrapers will be on that day, I do not know.

Tall Stories

Obviously, it is not possible within the scope of this chapter to provide descriptions of all the well-known big buildings of the world, but here are a few high-profile examples. Most of these tall structures (and many others not included, such as the CN Tower in Toronto) have an observation floor or a restaurant that is open to the public. Many have tours with knowledgeable guides, and all have their own websites, which contain technical information about the structure, its history, and sometimes even a web camera with live pictures.

Empire State Building

The Empire State Building is an American icon and a National Historic Landmark. Since it was completed in 1931, the Empire State Building has received over 110 million visitors to its "observatory" on the 86th floor, with panoramic views of New York. Constructed during the Depression, the "empty state building" (as it was dubbed when it opened because of the initial lack of renters during that troubled era) quickly became famous and easily recognized by many people at home and abroad, due to its distinctive

16. Concerns over terrorism may also limit the height to which institutions and nations are prepared to build, though I doubt it. Not that I think that terrorists will be always be prevented from striking a target, but now that we know tall buildings are targets, we can design them so that they can withstand any conventional explosion that a terrorist is capable of delivering.

FIGURE 3.11. The Empire State Building, New York. Image courtesy of John R. Plate.

Art Deco exterior. The setbacks on the exterior that provide the building's distinctive shape—seen in figure 3.11—resulted from zoning laws rather than from a desire for creating a distinctive style. The building was featured prominently in the original *King Kong* movie of 1933 and has appeared in many movies since.[17] It was the world's tallest building for nearly 40 years. Initially, it vied with its contemporary, the Chrysler Building, for this honor (much of the technical dispute over what is allowed to be included in the definition of building height dates from this period), and it held the crown

17. *An Affair to Remember*, *Sleepless in Seattle*, and *Independence Day*, among others.

until the North Tower of the World Trade Center was built in 1970. An eerie and unwelcome connection with the WTC Towers is the fact that both have been struck by airplanes. On July 26, 1945, a B-25 bomber accidentally crashed into the Empire State Building between the 79th and 80th floors. One of the aircraft engines shot out of the other side of the building, and another fell down an elevator shaft. Fourteen people were killed, but the building suffered no serious damage and reopened within two days.

Some basic statistics will convey to you the scale of the Empire State Building. It has 102 floors connected by 73 elevators and 1,860 steps (should you feel the need for a cardiovascular workout). The floor area totals 2,768,591 square feet, and the building weighs in at 370,000 tons. Twenty thousand employees work at over 1,000 different businesses within the building each working day. The steel frame supports stone cladding that includes 6,500 windows. No, I do not know how much it costs to clean all those windows or how long it takes, but I will be willing to bet that the window cleaners have no fear of heights.

World Trade Center Twin Towers

I will say something about the sad demise of the World Trade Center Towers in chapter 7; here, I will concentrate upon the structures themselves, in happier days when they helped to define the New York skyline (fig. 3.12). If they stood out from other New York skyscrapers, then this may be in part because their construction was not subject to New York City building codes but instead came under the umbrella of the Port Authority of New York and New Jersey. The towers were each 110 stories, though the North Tower was slightly taller. Indeed, it held the record for world's tallest building until that crown was taken by the Sears Building in Chicago. The South Tower had an observation deck.

The Twin Towers were completed in 1972 (North) and 1973 (South). They were the first tall towers to be built without masonry in their walls; each had a massive steel truss and concrete core, which took most of the weight, and a unique system of closely spaced outer steel columns (with a square cross section 14 inches on a side and 22 inches apart). The outer columns made the buildings stiff[18]—very resistant to wind loading—and permitted open interior space. These were load-bearing walls in that they

18. Recall from chapter 1 the stiffness of I-beams, hollow tubes, and other structures that have strong material on the outside.

FIGURE 3.12. The World Trade Center Twin Towers: (a) in happier times, defining the New York skyline; (b) the reduced skyline today. Again, I am grateful to John R. Plate for these photos.

supported the floors, which were formed from steel trusses spanning the space from walls to core. The floor trusses were of the deck, or underslung, Warren design shown in figure 2.8b, and were topped with concrete, which formed the floor surface.

The exterior tube skeleton permitted many small windows, and this design gave the Twin Towers a rather uniform appearance when seen from a distance. Early architecture critics referred to the towers as "the two boxes in which the Empire State Building and the Chrysler Building were packaged." Each "box" in fact contained about 3.8 million square feet of office space. Between them the two towers were the workplace for about 50,000 people and weighed in at about one million tons. One other connection with the Empire State Building: the Twin Towers were also a feature of a *King Kong* film, this time the 1976 remake.

Petronas Towers

Another set of twin towers, the Petronas Towers in Kuala Lumpur, Malaysia (named for the Malaysian state oil company), vies with the Sears Building in Chicago and Teipei101 in Taiwan as the world's tallest completed building. Each can lay claim to the crown based upon a narrow definition, but I won't go there; clearly, all four buildings are of similar height. I choose the Petronas Towers here, not for their impressive height but because of their double structure, connected by an unusual bridge, as you can see in figure 3.13.[19] This bridge, or "skyway," is the highest two-story bridge in the world and connects the 41st and 42nd floors of the two towers. (There are 88 floors in each tower, with a sophisticated elevator system between floors.) The engineer in you will have noticed the simple but gigantic truss structure that supports the skyway.

The foundations of the Petronas Towers are very deep (400 feet) because of the local geology and because the buildings are heavy: these towers are made from high-strength reinforced concrete, not steel as with most other modern tall buildings, and each is perhaps twice as heavy as a steel frame building of the same height. The tapering shape of the towers and the structure of the glass curtain walls are features redolent of Islamic architecture, appropriately enough in a largely Muslim country. The towers plus connecting bridge act as a gateway to the city and were built, of course, for the prestige that they confer upon a modernizing nation. As

19. Surprisingly, the two towers were built by different contractors.

FIGURE 3.13. The Petronas Towers, in Kuala Lumpur, Malaysia. I thank Peter Lee for this image.

with all tall buildings, the Petronas Towers are a tourist attraction and a magnet for movies (including *Entrapment*).

The Petronas towers were built sufficiently stiff that they require no special damping mechanisms to counter the effects of wind loading. Nevertheless, wind loading has to be taken into consideration for the skyway bridge that connects the two towers. This is because the towers will sway independently in a strong breeze—not very much, because they are stiff, but enough to damage the skyway if it is rigidly attached to both towers. The towers are so massive that if, for example, they were to move together a few inches at the 41st floor level, the skyway would buckle or would jam into the towers. So the designers made the skyway connections in such a manner that there is flexibility of movement at all points. The skyway connections are actually freer than the idealized truss pins of chapter 2, which, you may recall, permit free rotation. Here, the bridge endpoints can slide on rollers.

SUMMARY: In this chapter we have looked into the reasons that people want to build tall towers. We have applied the truss theory of chapter 1 to tall towers, beginning with the historically important (and architecturally decorative) buttresses that support high-status medieval towers such as those of Notre Dame Cathedral. We have also examined three impressive examples of modern tall buildings—the Empire State Building, the World Trade Center Towers, and the Petronas Towers—and have speculated how high such towers can go.

CHAPTER 4

Arches and Domes

Arch-itecture

We see arches all around us. The natural world carves out stone arches (as in fig. 4.1), and we photograph them because they are strange and yet familiar. The sculpting effects of wind-driven sand mimics the architects' construction; erosion will eventually bring down these natural structures, but the last remnant standing is amazing to us. We photograph it for the same reason we photograph arches that are standing defiantly in bomb-blasted cities: they seemingly shouldn't be there—but then, as everyone knows, arches are very stable.

Why is this so? In this chapter we will see why arches are stable using some basic physics (and a little math, for those who like math). We will look at different arches and at their three-dimensional equivalent: domes. Arches and domes have been a ubiquitous and integral part of stone construction for over 3,000 years, and many buildings made up of arches and domes, from the first millennium and earlier, are still with us today, even those that are located in earthquake zones. See figure 4.2 for two historical examples.

The Catenary and the Thin Arch

To understand domes, we need to first understand their simpler two-dimensional equivalent, the arch. The easiest way to understand arches is to consider an upside-down version, the hanging chain. This approach to appreciating the stability of arches and domes avoids some of the complex math that has been developed over the last three centuries by mathematicians, architects, engineers, and physicists to further our theoretical understanding. I can skirt the math and yet convey the essential physics of arches and domes because the reason for arch stability is basically geometrical.

Thus, it is quite easy to get across the key concepts with a diagram or two and a few paragraphs of explanation. This is the approach I will adopt here, and I begin by putting you in chains.

Well, not literally, but I will ask you to consider a chain problem that Jakob Bernoulli placed before his esteemed mathematical colleagues in the year 1690. During the last quarter of the seventeenth century Europe produced a large number of very able men (in those days women were only very rarely allowed to pursue scientific research), and they communicated their results to each other. Research journals sprang up to aid dissemination of new theories and experiments, and scientific societies arose at this time; of particular importance was the Académie des Sciences in Paris and the Royal Society in London. This networking—as we would call it today—of capable researchers helped to accelerate the progress of scientific understanding.

This was a period when mathematics was beginning to be appreciated as the language of Mother Nature, and philosophers (today we would call them physicists or mathematicians)[1] were developing math and embarking on the long process of applying it to nature—a process that continues to this day. The calculus was developed as a mathematical tool during the seventeenth century by two members of this eminent generation of philosophers: the Englishman Isaac Newton, as part of his quest to understand planetary motion, and independently by the German Gottfried Leibnitz. Newton and Leibnitz developed their versions of calculus using quite different notations, and something of both notations survives today, though generally we have adopted Leibnitz's.[2]

Jakob Bernoulli was a prominent member of a talented family of Swiss mathematicians who made a number of very important contributions. Jakob coined the term *integral*, as in integral calculus, and he first posed the problem that is of interest to us here. Consider a chain that is suspended between two points. Let us say that the end links of the chain are hooked over two nails hammered into a wall, at the same height. The chain is longer than the distance between the nails, and so the chain hangs loose,

1. The word *scientist* was not used until the early nineteenth century. Until then, the discipline that we call science was referred to as *natural philosophy*.

2. The two men engaged in a long-running and acrimonious argument about priority, over who was the first to invent the calculus. Unfortunately, this squabbling is part and parcel of scientific progress, given that scientists are people, and we will see more of it soon, in the small and pugnacious form of Robert Hooke.

FIGURE 4.1. Arches are stable structures, and natural arches can last for thousands of years. Thanks to Don Cooper for this image from the American West.

describing a curve between the two nails. An example of this chain-curve is shown in figure 4.3, using one of my wife's necklaces.

Bernoulli wanted to know how to describe this shape mathematically. Much applied mathematics in those early days was geometrical (indeed, Newton's version of the calculus was derived from geometrical considerations), and this is one example. Bernoulli challenged the other mathematical philosophers around Europe to find the mathematical function that described the chain curve, or *catenary*,[3] as it was called. Earlier, the great Italian scientist Galileo Galilei had wrongly claimed that the chain curve was described by a parabola. In fact, the difference between the actual chain curve and a parabolic curve is quite small, but it is there. Parabolas do play a role in the mathematics of structures, as we will see in chapter 5, but they ain't the chain curve. The German mathematician Jungius proved this, and a year after Bernoulli's challenge had been issued, in 1691, several philosophers had the right answer. Leibnitz, Johann Bernoulli (brother of

3. From the Latin *catena*, meaning "chain."

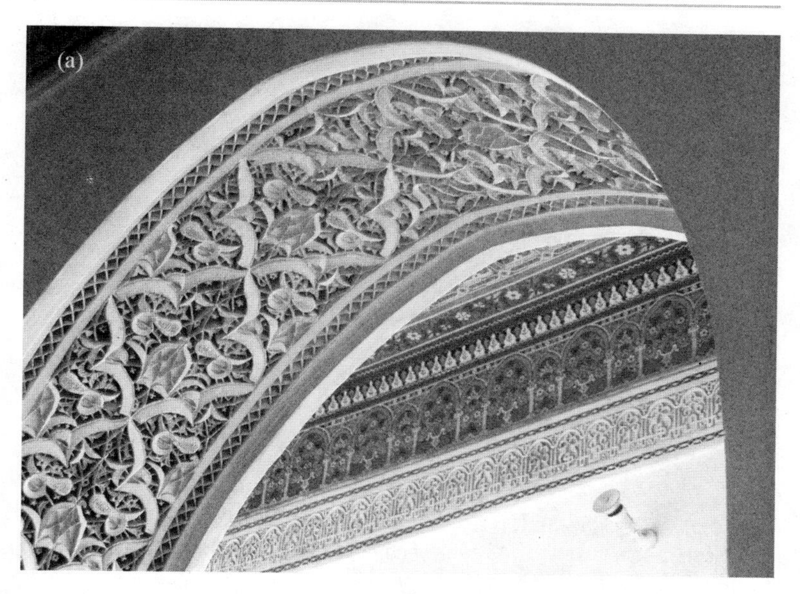

FIGURE 4.2. The arch has been a feature of masonry architecture for millennia. (a) An ornate example from Morocco. (b) The Blue Mosque, in Istanbul, Turkey, with its profusion of arches and domes. I thank Waldemar J. Poerner for these images.

FIGURE 4.3. A necklace that is suspended between two fixed end points describes the chain curve—the catenary. All free-hanging necklaces, chains, and ropes that are suspended between any two points form the same catenary shape. (Change the end points, and the catenary shape changes—but it is still a catenary.) At the end of the seventeenth century, Jakob Bernoulli asked the question, "How do we describe the shape of this curve mathematically?" Photo by author.

Jakob), and the great Dutch scientist Christiaan Huygens (who coined the name "catenary") showed that the chain curve was in fact a hyperbolic cosine. Those readers who crave mathematical details will find the catenary equation in note 6 of the appendix.

The catenary is a function that crops up in several mathematical places. It describes the curve of chains and necklaces exactly only for those that are infinitely thin. Real chains and real necklaces have finite thickness, and so the catenary is, strictly speaking, only an approximation to the true shape of the chain curve, but it is a very good approximation. Another restriction we must make upon the chain, if the catenary is to be a good description of its curve, is that the chain density per unit length must be constant. This is reasonable for a chain, and it is reasonable for the necklace of figure 4.3.

Some observations about the hanging chain will help when we come to considering arches. The chain is subjected to a tension along its length as well as a gravitational force that pulls it down. There is a stronger tensile stress on a chain that hangs with only a slight dip than on a looser chain; this is what we would expect. A chain (or rope or necklace) is floppy when not held in tension: it has no resistance to compressive force or to bending forces. This is the key to deriving the chain curve (and, perhaps surprisingly, to understanding how arches work). Because it has no resistance to compressive or bending forces, a hanging chain must adopt a shape that results in neither of these; if, say, a bending force acts on the chain, the chain will bend until there is no more force. Tension along the length of a hanging chain is a combination of the vertical gravitational force and the horizontal force that pulls the chain apart. So for an anchor chain, for example, the horizontal force is exerted by the anchor at one end and a ship at the other. The ship may be pulled by an ocean current away from the anchor, but the anchor pulls back to hold the ship; and the chain, caught in the middle, is subject to strong tensile force.

Here comes the connection with masonry arches. A thin stone arch is, to a physicist or engineer, mathematically very similar to an upside-down chain. Where the chain is strong under tension but weak when acted upon by other forces, the stone arch is strong under compression and weak when acted upon by other forces. So a thin stone arch (it must be thin and of uniform density per unit length, so that its mass distribution resembles that of a chain) must be made in the shape that produces only compressive forces along its length: any bending or tension forces will break the arch.

This idea may be made clearer by considering the arch shown in figure 4.4. The vertical force that acts upon any given block of stone in the arch is clear enough: gravity pulls the block downward. There is also a horizontal force. You can see from the figure that the block we are concentrating upon will be pushed to the right by the weight of the rest of the arch to its left. Yes, it is true that the weight of the arch to the right will also lean on our block and push it leftward, but because of our block's position it will feel more of a horizontal force pushing it towards the right. Now add (vectorially) this net horizontal force with the gravity force. If the resultant total force acts along the arch's length, then it is wholly compressive and the stone blocks can deal with it—the arch stands. If the total force acts in a different direction, so that the force vector points outside the arch, then the blocks that compose the arch will feel tensile or bending forces which they cannot resist, and the arch will collapse. An arch is said to be *funicular* if it has a shape that follows the direction of the lines of force, from top to bottom, (so that the vector for each stone block in fig. 4.4 lies within the arch). For a thin arch the funicular shape is an upside-down catenary.

The large curve of the Gateway Arch (fig. 4.5), an icon of St. Louis, Missouri, is very nearly an inverted catenary, with the parameter λ = 127.7 feet. It is gigantic—630 feet high—and is made from reinforced concrete with a stainless steel exterior. The arch, completed in 1965, is located by the Mississippi River in St. Louis, the "gateway to the West." There is a visitor center beneath the arch and a unique tram system that takes tourists up one arm of the arch to an observation area very near the top. The cross section of the arch is an equilateral triangle, but the size of the triangle changes with height. At the base each side is 54 feet long, whereas at the top of the arch each side of the triangle is only 17 feet long. The equation for a funicular arch with this varying cross section is not quite a catenary—recall that the catenary was derived assuming uniform density per unit length—but it is close enough. Whatever the exact funicular curve may be, it lies within the shape of the Gateway Arch, so you need not worry about it falling down as you gaze from an observation window near the top. In addition to being a technological and architectural achievement, the Gateway Arch is a testament to our understanding of geometry and forces.

We have seen that builders in ancient times learned how to construct empirically—by trial and error. This empirical learning extended, it seems, to the catenary arch. An ancient ruin in the Babylonian city of Ctesiphon

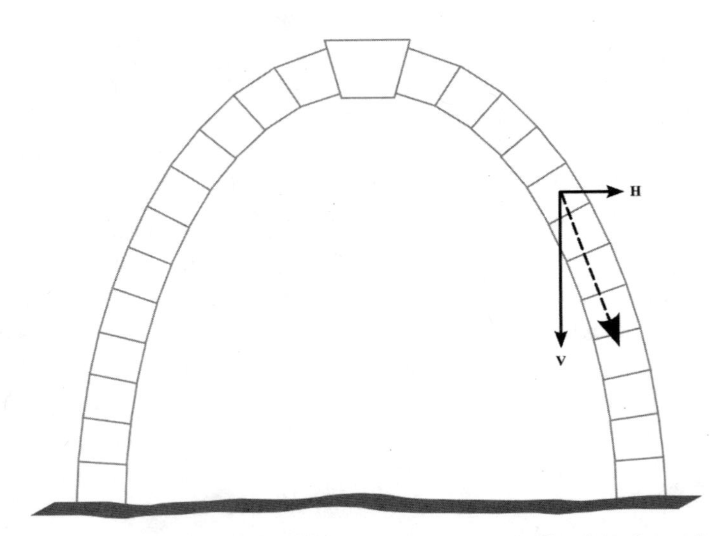

FIGURE 4.4. Each stone block in this thin arch carries a vertical load, V, due to its own weight and that of the blocks above it, and a horizontal load, H, due to the arch shape. For a catenary arch, the total force on each block (the vector sum of V and H, shown here for one block by the dashed arrow) is directed along the arch. Only compressive forces act on such an arch, so it is stable. An arch with a shape that follows the lines of force is called a *funicular* arch.

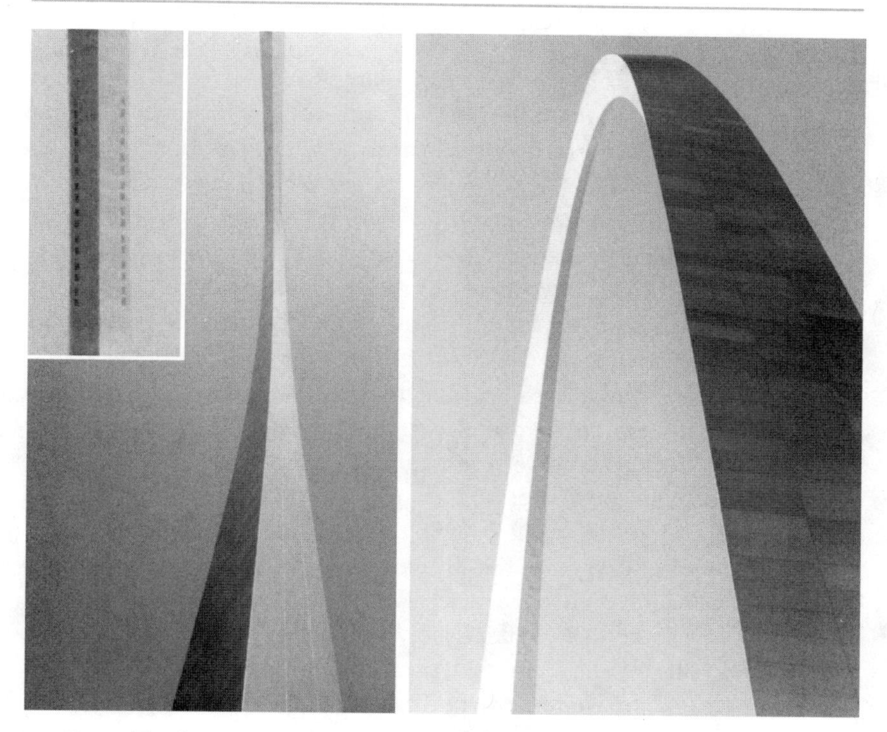

FIGURE 4.5. The Gateway Arch, St. Louis, Missouri. The triangular cross section and graceful catenary curve are evident. The inset shows observation windows near the apex. I am grateful to Marvin Berryman for these striking pictures.

(modern-day Al-Madaen, Iraq) dating from the first millennium BC displays a catenary arch. The top of this arch does not support any load other than its own weight. We have seen that the catenary is the funicular shape for thin arches that support only their own weight, and so the fact that this arch is still standing after more than 2,000 years tells us that it must be pretty close to the optimum shape.

Underneath the Arches

Most structural arches are not thin inverted catenaries, however. They carry loads that greatly exceed their own weight, and the funicular shape for such arch curves is not a catenary.[4] In this section I will show you how

4. The funicular shape for a load-bearing arch depends upon the load distribution. We will see an example in the next chapter.

arches evolved over the centuries, and how they became better and better at supporting heavy loads while spanning large spaces.

First, how does an arch differ from a beam? This is not such a simple question as it sounds; the answer has little to do with shape but depends on the way that the structure is supported. A beam obtains vertical support from the ground (obviously, since it does not fall through it), whereas an arch obtains both vertical and horizontal support. A diagram may help to explain this difference between beams and arches more clearly. In figure 4.6a you see a beam and a flat arch. The beam rests upon the ground and each end is supported by a vertical force from the ground. The flat arch is squeezed by the ground; a load passing over this flat arch would be transmitted diagonally, and so, if the arch does not break, then the ground must be reacting to the load with both vertical and horizontal forces. This is the key difference between beams and arches. If the flat arch of figure 4.6a touched the ground only at the bottom corners—let us say that the diagonal sides are separated from the ground by ⅛ inch—then the flat arch would not be an arch at all, but a beam. This would be true even if the stone block was curved like an arch: if it simply rests on top of the ground, then it is a beam. I will provide a simple truss example in the next section that shows the key role played by the horizontal component of ground force. Ultimately, the strength of the ground is utilized to oppose the outward thrust of the load. By harnessing this force the arch can support much greater loads than can the old post and lintel system and can span much larger spaces.

Arches first appeared in buildings at least 3,500 years ago in the Middle East. Egyptians, Babylonians, and Greeks used arches for secular structures such as sewers and granary storerooms, but they seem to have eschewed the arch for important visible buildings such as temples. For reasons that defy analysis, arches were considered to be unsuitable for monumental architecture. Many of the early arches were not true arches at all; they were *corbel*, or *false*, arches, shown in figure 4.6b. A corbel arch differs from a true arch (fig. 4.6c) in the manner in which it is built as well as in the load distribution. True masonry arches require a *centering* structure (wooden scaffolding that supports the stone blocks) before the last block—the *keystone*—is put into place. The arch becomes self-supporting only when the keystone connects the two curved arms. This factor of construction makes true arches more difficult to build than corbel arches, but they can span larger spaces. Intuitively we see that the keystone and the other wedge-shaped blocks (the *voussoirs*) near the top of the arch in figure

4.6c will push the walls outward. This is indeed the case, and so this arch may need *abutments*—buttressing at the sides, achieved with stone walls or mounds of earth. Most structural arches do require abutments, as we will see. On the other hand, the voussoirs of this particular arch are quite thick, and so the forces due to loading may be transmitted through the lowest blocks of the arch to the ground. Expressed differently, the funicular shape for this arch may lie entirely within the inner and outer curves defined by the stonework. If so, the arch is freestanding and stable.

The manner in which the load of an arch can be transmitted to the ground is suggested in figures 4.6d–f. The lines of force may pass through the abutments, along the voussoirs, or through the inside edge of the voussoirs.[5] In the last case the arch will collapse because, as we saw earlier, forces that are directed through the arch surface (except at the bottom) induce bending and tension forces, which masonry cannot withstand.

Historically, the Romans were the first people to use the arch widely, and they employed it as an architectural as well as a structural feature of their buildings. They generally observed the older tradition of avoiding arches for their religious buildings (the Pantheon being a striking and glorious exception), but the Romans built so much else, and so well, that many of their arched structures survive. Roman arches were semicircular, and they were used in buildings, bridges, aqueducts, and prestigious monumental edifices such as the Colosseum (fig. 4.7).[6] Masonry arches were the only alternative to the old post and lintel system until new building materials became available in the nineteenth century, and so arches had plenty of time to evolve. We have already seen why the arch became popular in most ancient cultures: it can carry a heavier load and span a wider space than the lintel. One other advantage, quite significant when many arches were to be part of a structure, is that arches require only small stone blocks, rather than gigantic lintels carved from a single block. This is important logistically because it was a lot easier to convey many stone blocks to a building site, and raise them, than to do the same for a few very large blocks. The

5. Over the centuries of masonry arch building a considerable technical vocabulary of arch terms has developed. In addition to *keystone*, *voussoir*, and *abutment*, you will find a plethora of other arch terms—*extrados, haunch, impost, intrados, soffit, spandrel*, and *springer*—defined in the glossary.

6. An elliptical amphitheater—the largest in the Roman Empire—this building dates from the first century AD and was the Roman equivalent of the Superdome (it could hold 50,000 cheering fans), though the sports played within it were deadlier. The Colosseum hosted gladiatorial and other games for 500 years.

Romans generally did not use mortar when piecing together their voussoir stones to form arches—they did not need to. The stones were dressed so as to fit together closely, and anyway, the large compressive load of masonry arches forced the stones together rather than pulling them apart.

Another type of arch, the *horseshoe* arch, first appeared in India, perhaps as early as the first century AD, and then migrated westward. It is also known as the *Moorish* arch, thanks to its adoption by the Byzantines (worthy architectural successors to the Romans) and the architects of early

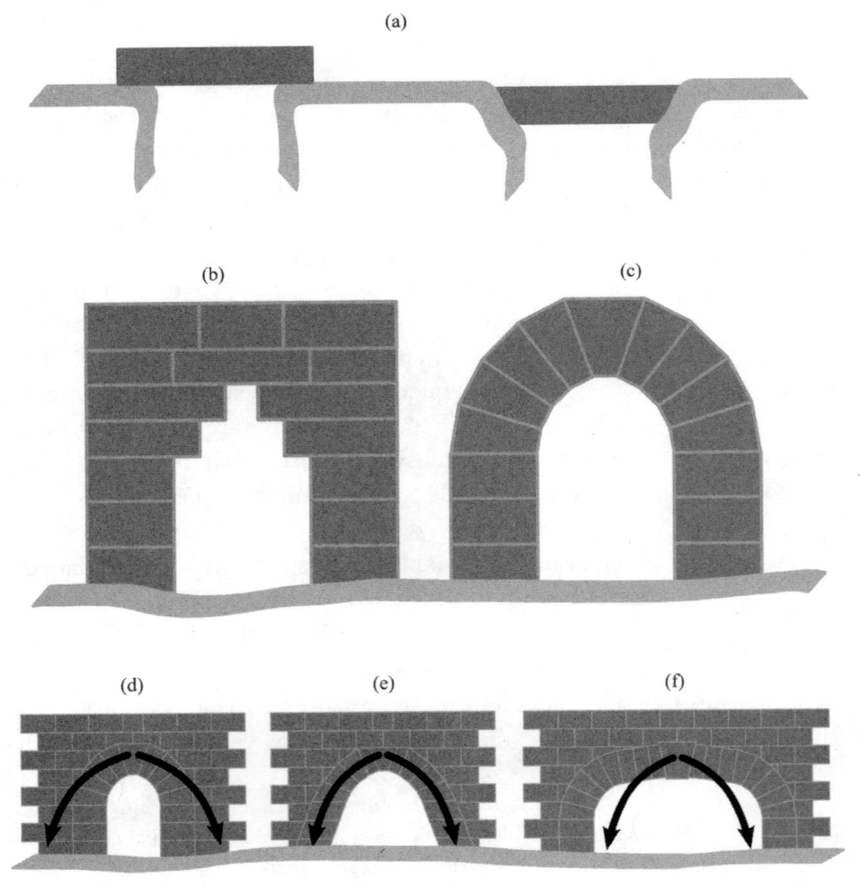

FIGURE 4.6. Beams and arches. (a) The stone block on the left is a beam, whereas that on the right may be an arch. (b) A corbel, or false, arch. (c) A true arch. (d) A stable arch with a shape that is not funicular. (e) A stable arch with a shape that is close to being funicular. (f) An unstable arch.

FIGURE 4.7. The Colosseum, Rome. Tiered arches were an icon of Roman structural engineering. This building was made of load-bearing masonry faced with cut stone. It has largely survived a number of earthquakes for nearly two millennia, but since the demise of the Roman Empire the Colosseum has been slowly but gradually stripped for its stone. Enough remains of its structural splendour to impress the thousands of people who visit it each year. I thank Waldemar J. Poerner for this image.

Islam. They developed and diversified the style of arches, and today, architectural historians can count many different types. The Moorish arch continues the curve of Roman arches beyond a semicircle (fig. 4.8). It became characteristic of Islamic architecture, being used both inside and outside buildings.

In western Europe, the builders of the Middle Ages preferred less than a semicircle, rather than more, resulting in either *segmental* arches or the *pointed*, or *Gothic*, arches that soar above many cathedral naves and transepts.[7] Pointed arches had the advantage over round arches that they required less buttressing because the horizontal thrust was less; we will soon see why. They were used architecturally to emphasize height. The repetition of fluted columns added more vertical lines to further enhance the

7. The pointed arch was in fact first used in Arab countries; returning Christian crusaders were happy to adopt it for their churches.

FIGURE 4.8. Islamic arches. (a) Moorish arch above a doorway in a mosque in Marrakesh, Morocco. (b) The eighth-century Mesquita mosque (a cathedral since the fifteenth-century reconquest) in Córdoba, Spain. This amazing structure creates internal space beneath a heavy ceiling by exuberant use of Moorish arches sitting on more than 1,000 slender columns. Thanks once again to Waldemar J. Poerner for these images.

impression of great height. These columns could be made quite thin, and this fact coupled with the arch height permitted larger windows; hence, more light flooded into the cathedral. "Raising the roof" was a common feature of medieval cathedrals, intended to inspire awe among worshipers (and also perhaps to make a statement about church power and prestige). From the engineering perspective, pointed arches are strongest at the apex. That is to say, a pointed arch can support a greater load at its apex than at any other point. We can see why quite easily by imagining the upside-down case: a hanging chain with extra weight suspended from its center will look like an inverted Gothic arch.

Many other styles of masonry arches emerged over the centuries—too many to describe here or even to define in the glossary. Most of these styles were chosen for visual effect and not for structural reasons. Two styles are worth further consideration: *elliptical* arches and the segmental arches mentioned above. These became popular more recently, in the early nineteenth century, when arch span became a more important consideration than arch height. The ellipse is a flattened circle, and elliptical arches were shaped like a semi-ellipse. Segmental arches were formed from segments of circles, so that they resembled flat, or jack, arches with just a slight curve. In both cases the horizontal thrust from the arch load was very large, for reasons we will now investigate; therefore, these arches required strong buttressing.

Truss Arch

Once again it will prove instructive to provide a simple truss model of a structure, the true analysis of which is far more complicated. As before, it turns out that our truss analysis, despite its simplicity, will yield insight into the forces that act within the real structure.

I might model the arch by a simple triangle, as in appendix note 3. We have already applied this triangle truss to a pitched church roof to show that if we removed the horizontal member, we must add buttressing to prevent the roof from collapsing. If we model an arch by a triangular truss and remove the horizontal member as before, the same reasoning applies.[8] Such a truss model teaches us two things about the forces that act upon

8. Such a structure resembles the three-pin arch, which we will discuss in chapter 5 when we look at large-span bridges. In olden times when structural materials were limited to stone and wood, arches were very important within buildings as well as for bridges. With the stronger and lighter materials available today, we can all but eliminate arches from buildings, but we still find that arches are useful for bridges.

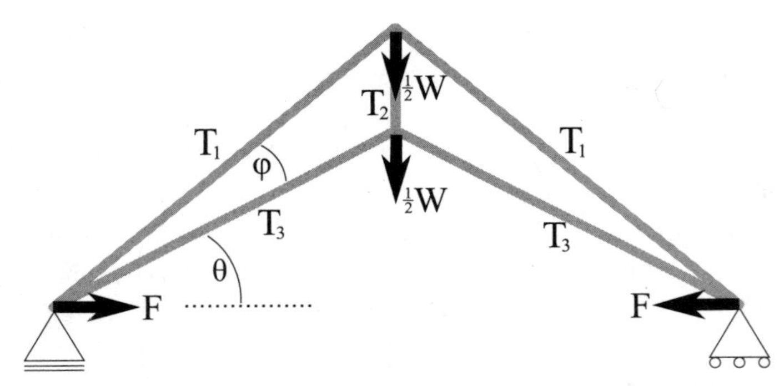

FIGURE 4.9. A truss model of an arch. The thickness and height of the arch are defined by the angles θ and φ. The truss load, *W*, is assumed to be split between the two central pins. Horizontal forces, *F*, at the ground represent arch buttressing. This truss is statically determinate, with eight equations and eight unknowns. Symmetry reduces this number. The horizontal ground reaction force is zero, and the vertical ground forces are $^1/_2 W$ at each of the two ground pins. The three remaining unknowns—the tension and compression forces T_{1-3} that act upon the five members—are discussed in note 7 of the appendix.

arches: for flat arches the compression loads are large, and the buttressing that is required to replace the horizontal truss member must be strong. Already we see why tall Gothic arches require relatively little buttressing and why flat, segmental, or elliptical arches require much buttressing.

In note 7 of the appendix, I construct a somewhat more complicated truss structure that tells us a bit more about the forces that act upon an arch (fig. 4.9). The analysis shows that for a masonry arch constructed of stone blocks or bricks (materials that are weak in tension), it is necessary to arrange the buttressing quite carefully; otherwise, one part or another of the arch will be in tension and so may collapse. If the buttress exerts too small a horizontal force, the arch may "do the splits," like a person with one leg on shore and one leg on a boat that is drifting. If the horizontal force is too large, the arch may be pushed inward, like cracking an egg by squeezing the sides too hard. Thus, the external buttressing that supports a masonry arch must be carefully designed to provide a quite specific horizontal compression that holds the arch together.

You might well ask what happens when a heavy load—such as snow on a roof—temporarily changes the normal loading of a masonry arch. If the buttressing forces are so tightly constrained, then is it not possible that such temporary loads might cause the arch to collapse? Well, with today's lightweight structures, it certainly is necessary to consider dynamical loading, as we will see in chapter 5. This was not a significant consideration for old-fashioned masonry structures, however, because the weight of the masonry (the static, or *dead*, load) was so much greater than that of any temporary load (the *live* load—literally so in the case of a passenger traversing a footbridge) that the live loads could be ignored.

Arcades, Arches, Vaults, and Viaducts

To some people an "arcade" brings to mind misspent youth amusing itself with rank upon rank of pinball machines and electronic games, while a "vault" is an event in Olympic gymnastics. To architects and structural engineers, though, these terms have specific meanings that are related to arches.

With the notable exception of the Gateway Arch, arches themselves are not usually massively huge constructions on the scale of the towers that we examined in the last chapter. However, arches are, and always have been, utilized in serried ranks to make big buildings—as the rows of arches of the Colosseum show (fig. 4.7). Arches can be strung together in two ways to form larger structures. First, they can be connected end to end, forming a long line. Roman aqueducts and Victorian railway viaducts crossing valley floors were made from such rows of arches, often tiered as in the Colosseum. It is not hard to see the advantages of building an aqueduct or a viaduct across a valley using arches. The amount of material needed to reach the desired height of the canal or railway line is reduced if the superstructure is formed from arches instead of solid stone or earthworks. Also, if the valley has a river running along the bottom of it, then the superstructure of, say, a viaduct will need to have gaps in it to permit water flow. The strength of a masonry viaduct built up from rows of arches is not much less than the strength of a solid wall of masonry—we saw in chapters 1 and 3 that the compressive strength of stone is so great that it can take almost any weight that we care to place upon it. Of course, the viaduct foundations must be firm, and the viaduct arches must be well designed so that the load is carried down the arch walls. What about abutments, which, we have seen, are required for all but the funicular arches? An arch that is one of a row becomes the required abutment for its neighbors. As a consequence, only the arches at the ends of a viaduct need external abutments, and these are provided by the valley walls. Thus, a row of arches forming a viaduct (or an aqueduct) crossing a valley floor can be very stable; the arches support each other and are held in place by the valley walls. Many

FIGURE 4.10. This arcade is part of a mausoleum in Tunisia. Each arch supports its neighbor, except where the arcade turns a corner. Note that the corner arches are provided with three columns for support. Thanks to Waldemar J. Poerner for this image.

Roman aqueducts and Victorian British viaducts survive today and are still functioning.

Rows of masonry arches were often used in neoclassical architecture to support the roofs of covered walkways. When the arches are placed side by side upon columns, as in figure 4.10, the resulting structure is known as an *arcade*. The use of arches and columns provides openness and light, which would be lacking if the roof was supported by walls. In Europe, many of the traditional university buildings were built around quadrangles; the covered walkways that formed the perimeter of these quadrangles were constructed as arcades. Students and professors could walk around the quadrangle and benefit from the air and light without getting soaked by rain. Monasteries, abbeys, and cathedrals were similarly cloistered.

There is a second way that arches can be strung together to extend a structure, and that is by placing them together so as to increase the depth of an arch, forming a tunnel. Architecturally, when the depth of an arch exceeds its span, we call the resulting structure a *vault*. When a vault is constructed, it requires centering or scaffolding, as do arches, until the keystones are in place. A row of arches (forming a vault) is not self-supporting in the way that a line of arches (forming an arcade or viaduct) is self-supporting. So vaults require external buttressing, unless the vault cross section is funicular. The simplest vault design is the *barrel vault*, which is an extended Roman or semicircular arch—half a cylinder. Such vaults have been around for as long as arches. The Babylonians constructed vaults out of clay bricks, and the Romans made extensive use of vaults constructed from stone blocks, both above and below ground, for storage and as sewers. One advantage of an underground vault is that the earth that covers the vault roof and walls provides the buttressing that is necessary to make the construction stable. One problem with above-ground vaults is that they admit no light; adding windows would compromise the vault's strength—it would be like removing one leg from an arch. (Light could be admitted by placing a vaulted ceiling on columns, perhaps forming an arcade, but care would be needed to provide abutments for the vault.) There are variations of barrel vaulting just as there are variations of Roman arches: two of these, segmental and pointed vaults, are illustrated in figure 4.11. There are many other variants, as for arches.

It would be wrong to think of vaults simply as stretched arches, however. The three-dimensional nature of vaults permits more structural variation than is possible for the two-dimensional arches.[9] One of the most significant structural additions that vaulting admits is the variation known as *groin vaulting*, also illustrated in figure 4.11. Groin vaults are two vaults intersecting at right angles (usually). The *groin* is the sharp surface of intersection (fig. 4.12). For intersecting barrel vaults of the same size, each

9. OK, OK, so arches are not really two-dimensional—they have depth as well as width and height—but from a structural engineer's point of view, arches can be analyzed as if they were two-dimensional. If we draw an arch on a piece of paper, the lines of force that arise from the load acting on the arch are directed downwards and outwards from the arch, but always in the same plane as the page (assuming that the arch is truly vertical).

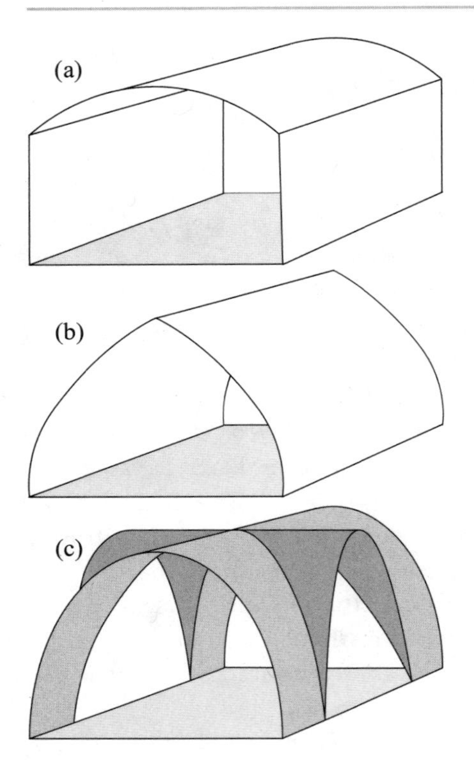

FIGURE 4.11. Arches become vaults: (a) segmental vault; (b) pointed vault; (c) groin vault.

with a semicircular cross section, the groin shape is a semi-ellipse. There are two main advantages of groin vaulting. First, light (and air) can be admitted to a main vault by adding several side vaults that intersect, with each side vault arching over a window. Second, for two intersecting vaults of the same size, the groin arches act as abutments for both vaults. In other words, the loading or thrust of the combination of vaults is directed down the groins, and so the perpendicular arches hold each other up, in the vicinity of the intersection. Thus, groin vaults are intrinsically stronger than barrel vaults (but more complex to construct).

The advantages of groin vaults were well known to the Romans, who spanned large spaces (up to 80 feet) with such structures, but full and widespread exploitation of groin vaults came in the churches and cathedrals built in Europe during the Middle Ages. The nave and transept of a church would often be vaults, and these intersected at right angles, providing mutual buttressing. Also, side vaults admitted light. Viewed from the inside by an awestruck worshipper looking to heaven, the high, airy, and light-filled Gothic cathedrals can appear to float on air, though they are solidly built of stone. This architectural formula for church design—with much use of arches and vaults—spread across medieval Europe and persisted for centuries. Even nineteenth- and twentieth-century cathedrals would be built on these traditional designs (fig. 4.13), though by the mid-1800s a new structural material, iron, had made the traditional masonry arch unnecessary.

Prior to the introduction of modern building materials, there was one other development of vaulted structures. We have seen how simple barrel

FIGURE 4.12. When two vaults intersect, the sharp-angled ridges that form are referred to as *groins*, and the combination of intersecting vaults is *groin vaulting*. Thanks to Antti Olavi Sarkilahti for this image.

FIGURE 4.13. St. Peter's Cathedral, Adelaide, Australia. The inspiring architecture of medieval Europe was later exported to the New World. Arches are everywhere: groin vaulting provides structural support and admits light, arcades allow easy movement between nave and aisles, and arches permit clerestory windows. I thank Brian Watmore for this image.

vaults were improved upon by groined vaults. Further improvement resulted from the innovation of *rib* or *ribbed* vaults, and of their derivative, *fan vaults*. Rib vaults first appeared in medieval times. At their simplest they consisted of ordinary vaults with some of the constituent arches made prominent, as shown in figure 4.14a. This feature made a vault stronger (imagine a truss being strengthened by replacing some members with stronger ones). Further improvement came with the addition of diagonal

FIGURE 4.14. Rib vaulting. (a) The simplest form of rib vault, with the supporting ribs perpendicular to the vault direction. (b) Rib vaults with cross ribs that stiffen the vault structure. These images are from the Mosteiro de Santa Maria de Alcobaça, Portugal—thanks to Antti Olavi Sarkilahti.

ribs, as shown in figure 4.14b (imagine a truss being strengthened by adding extra diagonal members). These arches would be constructed with buttressing so that they could stand freely, and the vault would be built around the arches. This means that the shape of the vault cross section (semicircular, pointed, or whatever) would be determined by the rib shape, and not the other way around. Rib vaults required less centering than barrel or groin vaults. Fan vaults took the rib idea to an extreme by concentrating one end of each rib arch around a single column. The result, popular in English perpendicular or Gothic cathedral architecture, is visually dramatic as well as structurally strengthening (see the fan vaulting shown in fig. I.2 of the introduction).

Domes

Like vaults, domes are arches that are continued into the third dimension. Instead of stretching an arch to form a vault, we rotate the arch about a vertical axis to make a dome. Unlike masonry arches and vaults, domes can be built with stone blocks without the need for centering (think of an igloo being constructed), and yet building domes is tricky, as we will see. A well-built masonry dome has almost every constituent block under compression and so is stable. It is a huge weight, and yet, if the builder was guided by an inspired architect with an eye for style, it could appear to hover in the air. In fact, the weight that is supported by the lowest row, or course, of stone blocks is less for a dome than it is for a vault.[10]

In this section we'll take a look at three famous buildings with gigantic domes, two ancient and one (relatively) modern: the Pantheon in Rome, Hagia Sophia in Istanbul, and St. Paul's Cathedral in London. Along the way we will see how the builders of these structures solved the problems of raising such domes, as well as take an excursion into dome geometry.

The Pantheon

The Pantheon in Rome was completed during the reign of the emperor Hadrian, in about 126 AD. The main part of the building consists of a

10. Think of a vault as a series of arches. Two opposite stone blocks at the foundation level support one arch. For a dome, two opposite blocks support an arch which becomes progressively thinner as it rises and which tapers to a point at the apex of the dome. So, in this sense, domes weigh less than vaults formed from the same arch.

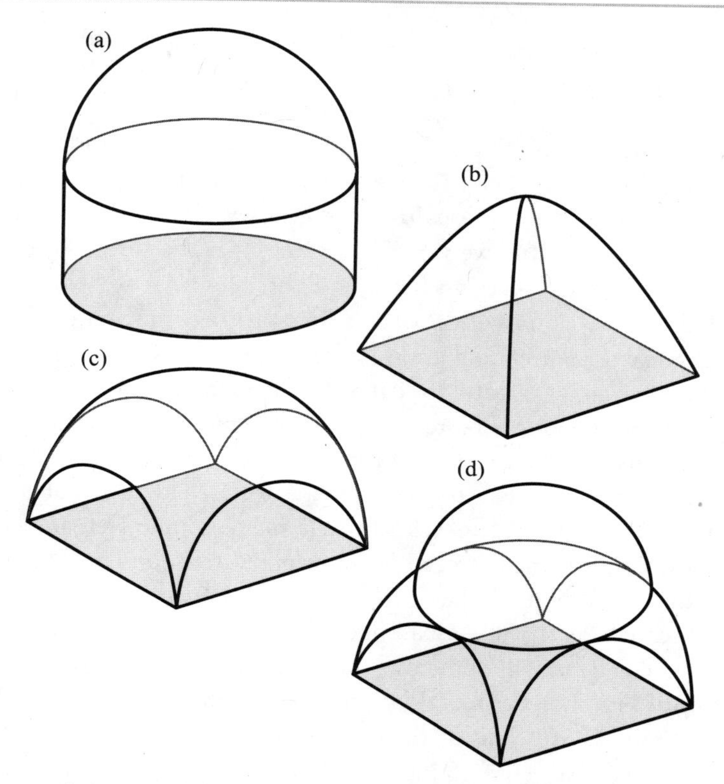

FIGURE 4.15. Four types of dome construction: (a) hemispherical dome on a cylindrical base—the Pantheon design; (b) cloister dome on a square base; (c) pendentive dome; (d) hemispherical dome on pendentives.

large hemispherical dome that sits atop a cylinder of the same radius, as sketched in figure 4.15(a). The inside diameter of the cylinder is 143 feet, and its height is half as much (so that a giant ball of diameter 143 feet could just fit inside the Pantheon). The only way that light can get into this huge open space is through the oculus, or giant eye, a circular opening right at the top of the dome. The oculus is 30 feet across, and so there is plenty of light inside.

A thought may strike you, upon reading about the oculus, about how different dome physics must be from arch physics. When building arches we needed to support the structure with centering, or scaffolding, which could not be removed until the keystone was in place at the arch apex. Here, we have a giant hole left at the equivalent place in a dome, and yet the structure stands. True, and you may also recall that I said domes need not require centering at all, depending upon construction technique. I do not want to get into the detailed modeling of dome forces (with a 3-D truss, for example), but I think by now that you may have developed an intuition for how it all works. Because a dome is three-dimensional, its weight pushes the voussoirs at the edge of the oculus together: they are in compression, held in a viselike grip by the weight of the dome. It is the three-dimensional nature of domes that accounts for their differences from arches. One similarity: the force that pushes the oculus voussoirs together would be greater if the dome were flatter, and less if it were taller, just as the lateral force at the apex of an arch is greater for flat arches than for tall ones.

To allow for the lines of force through the dome (to ensure that the 3-D funicular curve is contained within the structure) the thickness is greatest near the base of the Pantheon's dome: 21feet. Near the oculus, dome thickness decreases to 4 feet. To further reduce dome weight, the concrete of which it is made (recall from chapter 1 that Romans knew about concrete) is less dense.[11] The cylindrical wall beneath the dome is the same thickness as the dome base, but the wall is stonework rather than concrete. The Pantheon dome was the largest in the world for over 1,300 years until Florence Cathedral was built in the fifteenth century. Today, the Pantheon is still one of the best preserved of Roman buildings, a testament to the strength of domes and to the skills of its unknown architects and builders.

Dome Geometry

It is usually considered desirable for buildings to have a square or rectangular floor plan, for all kinds of practical reasons. The Pantheon is unusual in its circular plan, but there is good reason for making the walls that support a dome circular. How did the builders of classical antiquity solve the problem of placing a dome upon a *square* base? There are a number of ways in which this can be done. Small domes can be constructed as in figure 4.15b. Such "cloister" domes have a square cross section when sliced horizontally and a curved arch shape when sliced vertically.

Another and more elegant solution is shown in figure 4.15c. Please regard the following recipe for placing a dome upon a square base as a piece

11. Pumice was used for the aggregate constituent of the lighter concrete. Even with this light concrete, the dome weighs 5,000 tons.

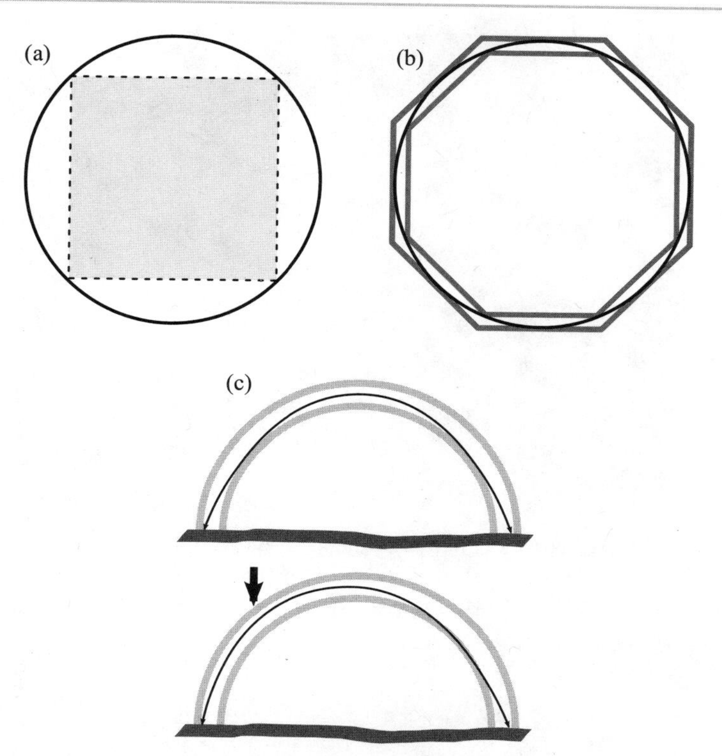

FIGURE 4.16. Dome geometry. (a) A dome can be placed upon a square base geometrically as shown here (plan view). Make the dome diameter equal to the diagonal, and cut the dome vertically along the dotted lines. The result is a pendentive dome, as shown in fig. 4.15c. (b) It may be easier to put the dome upon a different polygonal base, here an octagon. (c) Without abutments, a dome must be thick enough so that the load is directed through the walls and base. If there is an asymmetrical load then the structure must be able to deal with it. Herein lies a difference between thin arches (and domes) and hanging chains.

of geometrical construction, rather than as practical building instructions. Place a hemispherical dome upon a square base of such size that the corners of the square just touch the circular base of the dome (the plan view shown in fig. 4.16a may help you visualize this). Trim the parts of the dome that overhang the square with vertical cuts (just the way my mother would trim excess pastry off a pie crust); what remains will look like figure 4.15c. *Voilà*—a dome upon a square base. What's more, your dome lets in light without letting in rain. You can raise the dome to greater height by further geometric manipulations. Slice off the top of the dome and stick on

FIGURE 4.17. Decorated pendentive (triangular section of a sphere) from the National Library of Finland. Four such pendentives form the transition from circular dome to square base. I am grateful to Antti Olavi Sarkilahti for this image.

FIGURE 4.18. A ribbed dome upon a polygonal base; this example is from Tunisia. Thanks to Waldemar J. Poerner.

a smaller hemisphere, as suggested in figure 4.15d. Here you have the basic design of Hagia Sophia, of which more anon.

The spherical triangles that support the rest of the dome in figures 4.15c,d are known as *pendentives* (an example is shown in fig. 4.17). In practice, for real masonry structures, these pendentives are large stone constructions that support a massive vertical load. If you wish to build a dome without such concentrated loads, then you might consider changing the base from a square to some other polygon, such as the octagon shown in figure 4.16b. Clearly, the larger the number of sides that you choose for your polygon, the closer you are going to approximate the circular base of the dome it supports. Many domed buildings, particularly in the Islamic world, were constructed in this way (see fig. 4.18 for an elegant example).

We have seen that masonry arches generally require abutments to prevent them from falling. Similarly, domes will push out at the base and then collapse unless buttressing is positioned around the base to counter these lateral forces. Some old buildings were provided with a ring of iron chains at the base. Chains are strong in tension, and this method worked fine for small domes. For the large structures of interest to us in this book, the sideways forces are too great for chains to hold them, and so they must be countered in some other way. Some masonry domes could be made sufficiently thick[12] so that buttressing was not necessary (fig. 4.16c), but generally it was required. The somewhat ungainly, lumpy exterior of Hagia Sophia is due to the very large abutments that were needed to hold up its very heavy dome.

Hagia Sophia

Hagia Sophia was constructed in Constantinople during the reign of the greatest Byzantine emperor, Justinian. We know the names of the engineer-architects who designed and oversaw the building (in just 5 years!) of this magnificent structure—the most impressive building that I have ever seen. Anthemios and Isidoros were prolific architect-engineers within Byzantium, and they have left a legacy that extends beyond Hagia Sophia; the traditional church design throughout the Orthodox world—Slavic as well as Greek—is theirs. Hagia Sophia, however, marks the pinnacle of their achievements. Completed in 537 AD, this "Church of Holy Wisdom" was

12. The radius of a classical masonry dome would typically be 50 times the dome thickness. We will see that this ratio is much higher for modern domes.

FIGURE 4.19. Hagia Sophia, in Istanbul, Turkey. (a) An old drawing (from Wikipedia) showing the bulky exterior half-domes and other buttressing needed to support the large dome, which is raised on columns. (b) The dome is on the left and one of the half-dome buttresses is on the right—separated by a massive pendentive. This design created a huge internal space. (c) One of the arch colonnade "side corridors" conveys the scale of this massive structure. Thanks to Waldemar J. Poerner for the two photos.

built on a massive scale: it was the largest cathedral in the world at the time and would retain this record for 900 years, until the cathedral at Seville in Spain was constructed. The large dome of Hagia Sophia (107 feet in diameter, with the apex 183 feet above the floor) is not quite as big as that of the Pantheon, but it is raised on pendentives (the first time that this Byzantine innovation was used) and is then further raised up on an unbroken arcade of 40 arched columns that permit windows, and thus light, into the enormous interior space, as you can see from figure 4.19. The designers were clearly more concerned about creating an impressive interior than an imposing exterior, and much of the complex structure of the building is designed to create a large interior space. You can see something of the polished marble pillars and wall coverings in figure 4.19.

Hagia Sophia's first dome collapsed in 563 AD following an earthquake and was replaced.[13] The new dome was 20 feet higher than the old one, and this difference reduced the outward thrust of the dome, thus relieving the load upon the buttressing. More repairs were required in the ninth and fourteenth centuries, again following earthquake damage. The Ottoman Turks captured Constantinople (today's Istanbul) in 1453 AD. A century later the great Ottoman architect Sinan directed engineering works that further strengthened the massive building, which after the conquest had become a mosque (it is a museum today).[14]

Robert Hooke and St. Paul's Cathedral

We can conveniently place the scientific understanding of structural engineering concepts like arches, vaults, and domes in the seventeenth century, with a realization by Robert Hooke that showed an application of scientific thinking to the building of these structures. But while running a finger along the dateline of history and pausing in the 1670s is convenient, it is not by any means the whole story, as we have seen. Domes were appreciated two millennia before the combative Hooke fought his way

13. Yes, Hagia Sophia has survived for one and a half millennia despite being in an earthquake zone.

14. Hagia Sophia inspired Muslim Ottoman architects to build several mosques in Istanbul that could rival the Byzantine building for grandeur. With the Blue Mosque and others they succeeded. One of the most impressive features of Hagia Sophia, for me, is the fact that it was built 1,000 years before most of the rest of the world—Christian or Muslim—learned how to create grandeur on such a colossal scale.

FIGURE 4.20. A splendid example of a traditional masonry dome, this one in Paris. The successors to Rome in the east, the Byzantines, taught us how to place domes upon columns, so that light could flood inside. Thanks to Don Cooper for this image.

through life, and some of the largest and most magnificent domed structures were already 1,000 years old by that time. The designers and builders of these great buildings knew no modern science but were propelled by experience and an empirical approach that led to incremental improvements over centuries, as father taught son. Domes securely held their place in art as well as architecture (fig. 4.20) and were, in the late 1600s, much more secure than the science which later would revolutionize dome construction. I will highlight Hooke's contribution here; the application of scientific thinking without too much math is a feature of his long and productive life and is a feature of this book, so the story is worth telling.

"The single greatest experimental scientist of the seventeenth century." So a recent account of Robert Hooke describes this bona fide polyglot genius. A diminutive man with a curved spine, sickly for most of his life, Hooke was difficult and cantankerous. He famously got into acrimonious disagreements with many of the well-known and influential scientists of his day, which would rumble on for years, and these spats damaged Hooke's reputation for 200 years. Over the last few decades he has been the subject of several increasingly sympathetic biographical accounts. I mention him in this book not because he was a great engineer or a very competent architect (though he was both); there were architects before him and since who are more famous, and justly so. Rather, I highlight Hooke here because he was the person who made the simple yet insightful observation about inverting chains to obtain arches. This book aims to provide its readers with a (mostly) nonmathematical insight into big structures and why they are built the way that they are; Robert Hooke epitomizes nonmathematical insight, and then some.

Hooke made significant contributions to many fields: chemistry, microscopy, geology, meteorology, medicine, architecture, celestial mechanics, optics, horology, navigation, and not least, surveying. In addition, he was curator of experiments to the new Royal Society, a linguist (he wrote at least four languages), and a prolific inventor: the universal joint, the spirit level, the iris diaphragm, the sash window, and the weather station are among the products of his fertile mind. Hooke claimed to have been the first person to suggest that gravity obeyed an inverse square law. This may have been true, but he could not provide a mathematical proof, which Newton could, and so began a long and vindictive feud. Hooke claimed to have invented the anchor escapement mechanism, which had recently revolutionized horology, as well as the balance wheel and the spring assembly. Both these claims brought him into conflict with the great Dutch scientist Christiaan Huygens. He disagreed with Hevelius over astronomical observations and with Newton again over the nature of color. You see what I mean about his being difficult?

Hooke is remembered today by physicists and engineers because of Hooke's law of elasticity, which we encountered in chapter 1. He is remembered by astronomers through a couple of craters named after him, on the moon and Mars, and by readers of this book because of his advice to Sir Christopher Wren about how best to build domes.

London was decimated by bubonic plague in the years 1665–66.[15] This incidence of plague was killed off, along with the fleas that most likely spread it and the rats on the backs of which the fleas lived, by the Great Fire of London, which began in a baker's shop in September 1666. The conflagration raged for five days and nights, consuming most of the old city

15. To be pedantic, London was decimated twice over, because it lost a fifth of its population, rather than a tenth, as the term *decimation* implies.

and rendering most Londoners homeless, though mercifully, few were killed. Plans to rebuild the city were quickly put into place. Robert Hooke was appointed as the city surveyor. His friend and fellow scientist Christopher Wren was commissioned to design a new St. Paul's Cathedral on the site of the old one.[16] Wren's design was eventually accepted, though he modified it a lot over the 35 years it took to complete the rebuilding. One of the modifications was to the dome, and these modifications resulted from some advice that Wren received from Hooke.

Insight through physical intuition, without math: a thin arch is physically similar to an inverted hanging chain. Tension becomes compression but the same stability criterion exists: the lines of force must follow the curve. So, to be stable, thin arches must be catenaries. Rotate a thin arch and you get a catenary dome, which requires no buttressing.[17] All other masonry domes (of uniform thickness) require buttressing of some sort. Wren wanted a hemispherical dome for St. Paul's, and Hooke suggested to him how it should be built. Between the two of them emerged a plan for a triple dome, nested like Russian dolls, and Hooke knew how these were to be constructed. The lower dome is visible from the interior and is made of

16. The new version, the fourth, still stands today. The original had been constructed 1,000 years earlier.

17. This is not *quite* true, for the reason we saw earlier in note 10: the stress upon dome masonry is less than that acting on the corresponding arch masonry. However, for thin domes the difference between the true funicular dome surface and a rotated inverted catenary is very small. An indication of the increasing application of mathematics to science and technology of these times is provided by Giovanni Poleni, 50 years after Hooke. Poleni was an Italian mathematician who generalized and made more sophisticated the connection between hanging chains and arches. When considering the dome of St. Peter's in Rome—it had cracked and needed repairing—he considered first a chain hanging with different weights attached at different points, reflecting the variations in thickness of the dome, and so was able to estimate the shape of the funicular surface.

The response of thin arches and domes to a *changing* load highlights a difference between these compressive structures and their upside-down equivalent, the tensile hanging chain. Place a temporary load upon a hanging chain (say a pedestrian walking across a chain or rope bridge) and the chain shape changes to accommodate the load. Place a temporary (heavy) load upon a thin arch or dome and it breaks. Stone is brittle—it cannot flex. So, Hooke's insight is valid only up to a point: arches and chains are not perfect reflections of one another. In practice, masonry structures were generally so much heavier than any temporary load that builders could ignore the problem. When a dome weighs 5,000 tons, there is no need to worry about someone walking across the roof.

wood, to save weight. The heavy upper dome is visible from the outside and is supported by an invisible central dome that has the same base but is of a peculiar conical shape with a rounded top. (The central cone also supports the ornamental lantern structure seen atop the outer dome.) This middle dome is built of brick and is connected to the outer dome by timber frame scaffolding. Odd, like Hooke, but the design has lasted 300 years, and it survived the Blitz, despite being targeted by German bombers and hit by bombs.

Modern Mathematical Roof Structures

The masonry arches, vaults, and domes of classical and Renaissance Europe are reflected in our modern equivalents. The structures of today serve the same function as their ancestors—they cover spaces while providing large, open interiors that are protected from the weather, and they look good—but the family resemblance is only skin deep. Modern arches, vaults, and domes are, to a structural engineer, very different beasts from their Roman or Renaissance grandparents. We have seen several times in this book that the form of stone-built structures is constrained by the fact that stone is weak in tension, and so, for example, masonry arches and domes needed abutments to ensure that no part of the structure was subjected to tensile forces. The advent of new building materials in the nineteenth century, and also our better understanding of the forces that act upon large structures over the last century, has changed everything.

Cast iron and then steel beams, prestressed concrete, and steel cables enable us to construct arches, vaults, and domes that require no buttressing. A hemispherical reinforced concrete dome can be placed upon a base without concern about lateral forces: only the weight of the dome needs to be considered. The dome structure itself is strong enough to handle the forces that arise from its own weight. Furthermore, that weight is less than the weight of masonry domes because we can build domes that are much thinner than their earlier counterparts. Where a masonry dome would have a ratio of radius to thickness of 50, typically, a modern dome can attain 800. As a consequence, we are now capable of covering much greater areas with our vaults and domes, and even of making them movable, as millions of people have experienced at many of today's major sporting venues.

New understanding also contributes to the freedom of design that modern architects enjoy compared with their classical counterparts. We have seen how mathematical advances led to hyperboloid towers—curved shapes built up from straight components. Roof coverings have an exact equivalent, called *hypar*, which is short for "hyperbolic paraboloid." Large roofs of some public buildings (cathedrals,[18] campus buildings) are made in the hypar shape from steel beams and panels or thin reinforced concrete, while more modest, lightweight hypar roofs are constructed from steel cables and canvas. The idea behind these gracefully curved roofs is shown in figure 4.21.

One characteristic of hypar roofs (sometimes called "saddle roofs," for reasons that a glance at the figure makes obvious) is that the forces arising from a uniform load—due to snow, for example—are the same all across the roof. Whether the forces are compression (in the downward-curved parts of the roof) or tension (in the upward curved parts), the magnitude of the force is the same under uniform load. This characteristic means that hypar roofs can be constructed with less of a safety factor than other roof designs (which can concentrate loads at particular points) and so can be relatively lightweight. Further, the distribution of compressive and tensile forces acts to make the overall structure stiff, so a reinforced concrete hypar roof can be thinner than most roofs covering the same area. Hypar roofs are only infrequently used for domestic structures, for the very good reason that they are expensive. One example: curved concrete requires tailor-made forms. So, striking as they are, these curved roof surfaces made from straight lines will probably not be over your head as I write.

Another modern roof design, one more obviously related to earlier designs, is the geodesic dome. Geodesic domes are associated with the American engineer and architect Buckminster Fuller, who designed many such structures and who patented the idea in the United States, although he did not invent these domes.[19] Like hypar roofs, these designs depend upon modern building materials and are light and strong for their weight. Unlike the hypar shape, however, geodesic domes do at least have a superficial similarity to classic dome roofs. Geodesic domes are not necessarily spherical although most of them do approximate a sphere made up of many triangles. Triangular trusses are strong, and the combination of triangles to form curved three-dimensional structures is mathematically quite old

18. The Gaudi-inspired Sagrada Familia basilica in Barcelona, when eventually it is finished, will contain both hyperboloid and hypar structures, a favorite of this extraordinary early-twentieth-century Catalan architect.

19. The most famous geodesic dome is a Fuller design: the 20-story dome that formed the American exhibit at Expo 67 in Montreal.

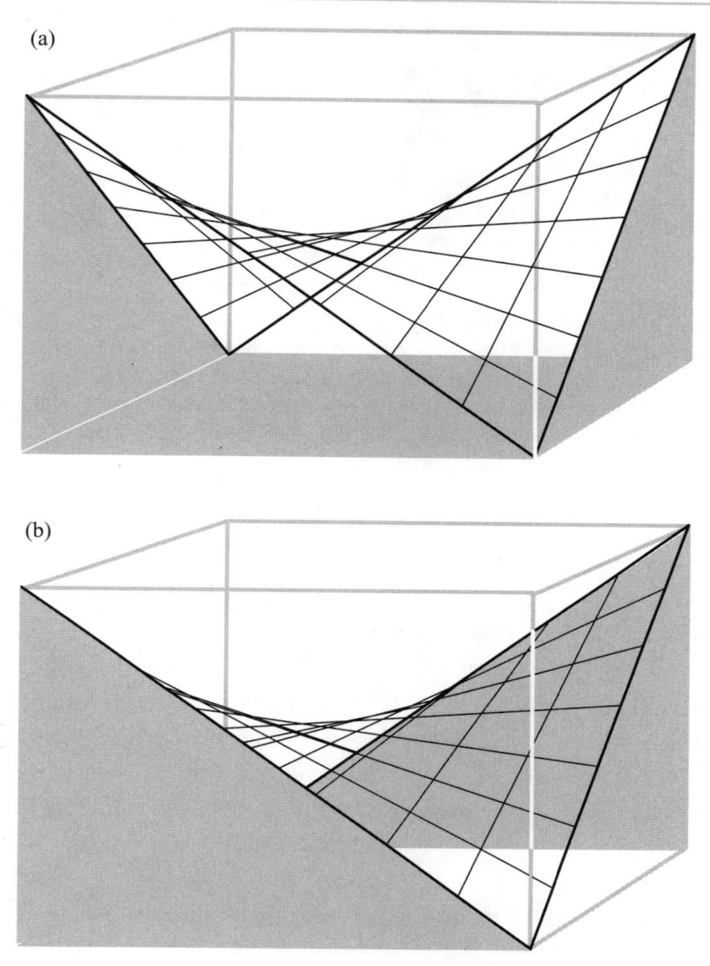

FIGURE 4.21. A hypar roof structure created from straight beams or taut cables (black lines). The roofed space consists of a square floor and four triangular walls. In (a) the left and right walls and the floor are shaded. In (b) the front and back walls are shaded. Several of these "butterfly" roofs can be combined to form more complex structures.

but architecturally a child of the twentieth century. First the math: a simple example of geometrical construction that leads to a geodesic dome is shown in figure 4.22. There are many, many ways that triangles (and other polygons such as pentagons and hexagons used in combination—for example in a soccer ball) can be combined to form a sphere or part of a sphere.

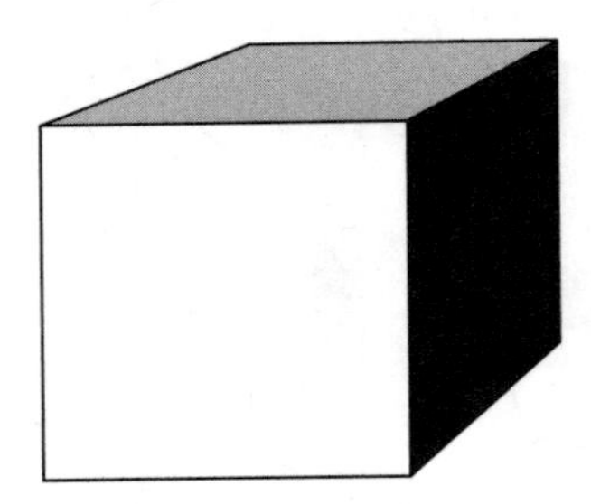

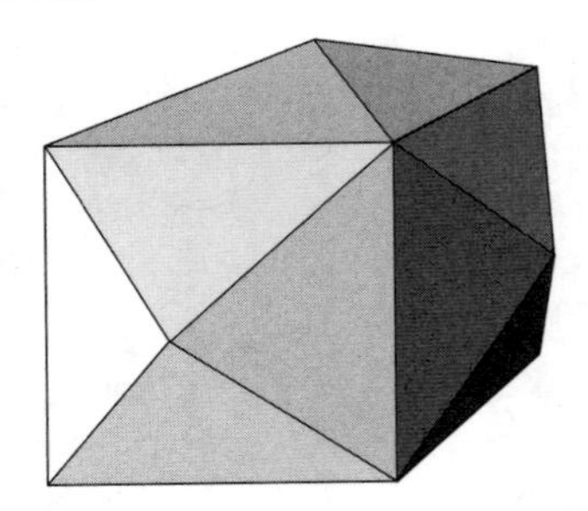

FIGURE 4.22. Geodesic structures. Here, five sides of a cube (four square walls and a ceiling) are each extended into pyramids, creating a geodesic structure consisting of 20 triangles. Each of these triangles could be extended into tetrahedra, leading to a structure that is composed of 60 smaller triangles and that more nearly resembles a sphere. The original cube could be replaced by a dodecahedron or by some other regular solid. The possibilities are endless.

Stated a little more carefully: the tubular steel members form a skeleton that approximates a sphere.

These lightweight truss structures are largely in tension. Consequently, their development could not even have been contemplated until high-tensile-strength materials became available. The polyhedral shape distributes forces within the structure; as is the case with reinforced concrete domes, there are no external forces required. Sometimes the term *tensegrity* (from "tensile integrity") is used to describe the principles involved in constructing strong and stable structures out of three-dimensional patterns of mutually reinforcing triangles.

All these modern thin-shell structures rely upon curvature to distribute loads, and they are very efficient at resisting distributed loads due to snow, wind or dead weight. They are less good at resisting concentrated point loads,[20] and the analysis of their response to point loads is complex. For truss-based structures such as geodesic domes, the complications are not too bad—a computer can readily handle the truss analysis—but for solid shell structures such as reinforced concrete hypar roofs, the analysis is not

20. An example commonly trotted out to illustrate this point is the egg. This shell structure easily resists its own dead weight and can in fact resist significantly larger distributed loads. However, as you demonstrate every time you crack a boiled egg for breakfast, the egg shell is vulnerable to concentrated loads. Egg shells, incidentally, have a ratio of span to thickness that is closer to that of classical domes than to that of modern domes (about 50:1 or 100:1) and for very good reason. Eggshell is brittle; like stonework, it is strong in compression but not in tension.

at all easy and is still an area of active research. Imagine the extra complexity involved in analyzing and building very large modern steel roof structures, such as those of the Superdome in New Orleans or the Bird's Nest stadium in Beijing.

Classical and neoclassical architecture is unthinkable without arches and their derivative structures, vaults and domes.

SUMMARY: The stability of arches is apparent in the durability of naturally formed arches. The shape of a thin arch—one that supports only its own weight—is the inverted shape of a hanging chain. Load-bearing arches can have different shapes. A simple truss construction has given us insight into the forces that act upon an arch. In particular, arches give rise to horizontal forces that must be resisted by abutments; the strength of these forces depends upon the shape of the arch and the load that it carries. Arches are a very important aspect of classical architecture—of its beauty as well as its form—as are the arch's three-dimensional extensions: arcades, vaults, and domes. Modern building materials permit different structures that replace the dome. Hypar and geodesic structures are subjected to different forces than those that arise from classical arches and domes.

CHAPTER 5

A Bridge Too Far

Spanning the Globe

We examined bridges of a limited kind in chapter 2, where truss theory was our main focus and truss bridges provided good illustrative examples. The bridges I described there were of an elementary type; our truss structures replaced simple beams. In chapter 4 we examined the various ways in which internal space can be created inside buildings, the overarching theme being, well, arches. I managed to get through that chapter without saying much about bridges, though nowadays this is where we see most of our arches. Beams and arches constitute the two traditional types of bridges, but as you can see in figures 5.1 and 5.2, many other types exist today. Why the explosion in bridge evolution? Two reasons: first, new materials have arisen over the last 200 years, leading to new design possibilities, and second, the ambition of bridge builders has increased in step with technology. Nowadays, there is a need for longer bridges, which the new technology and an improved technical understanding produce. Longer bridge spans mean that there are more sites where bridges can be built, and so the number as well as the size of bridges has increased dramatically, particularly in the past 80 years or so.

As bridges became bigger, they became icons—a common theme of the big structures in this book—and they conveyed prestige upon the builder and the city or state that paid the high and spiraling cost of construction. Perhaps for this reason we tend to remember the designer or the chief engineer who was the driving force behind a particular big bridge. We may not remember the name of the man who built the Empire State Building,[1]

1. There may even be a few cerebrally challenged people out there who do not remember the name of the person who designed the Eiffel Tower, but none of my readers, I hope.

but often we do know the human story behind a bridge construction: Roebling and the Brooklyn Bridge, or Strauss and the Golden Gate Bridge, for example. Perhaps this is because big bridges stand alone in very visible places, and we can see every aspect of their construction when they are being built, and of their structure once completed. We can appreciate the enormous effort and expense that was needed for the construction, and gawping at the breathtaking span, we can readily appreciate also the courage of the designer.

A glance at the numbers in figure 5.3 will show you how bridge spans have increased to barely believable lengths over the last century. The graphs in this figure also tell us that suspension bridges have held the records for span length for 80 years. (Later, I will estimate the maximum span attainable for a steel suspension bridge.) There are sound structural reasons why this is so, and in this chapter we will see why. The newest class of bridge design, cable-stayed bridges, come in (a distant) second in achieved span. You may ask, Why bother with a new design if it cannot do better than suspension bridges? Read on. Steel truss bridges, cantilever bridges, and arch bridges trail behind, and yet truss bridges and arch bridges are still being built. Why, and why not cantilever bridges? Here, the physics behind these different bridge designs is revealed, and their advantages and disadvantages are made clear.

Beam Bridges

The oldest type of bridge is, of course, the simple beam—a tree trunk across a stream, for example. I have discussed beams in chapter 1, and so we are already familiar with their characteristics. To be big, a beam bridge must be very strong in tension, and historically, the only bridge building material that was strong in tension was wood. Thus, for a very long time beam bridge span was limited by tree trunk length. Today, steel replaces tree trunks. The longest span for a beam bridge is 420 feet (an I-beam girder road bridge in Louisiana). Virtually the sole advantage of beam bridges, apart from the simplicity of design and construction, is the fact that they require only vertical piers to support them. There are no horizontal forces generated, and so abutments are not needed. End of story: we are interested here in *very* big bridges, and beams won't do it for us. I will take the main bridge design categories roughly in chronological order, which mostly corresponds to increasing bridge span. Let us quickly move on to the more sophisticated arch bridges.

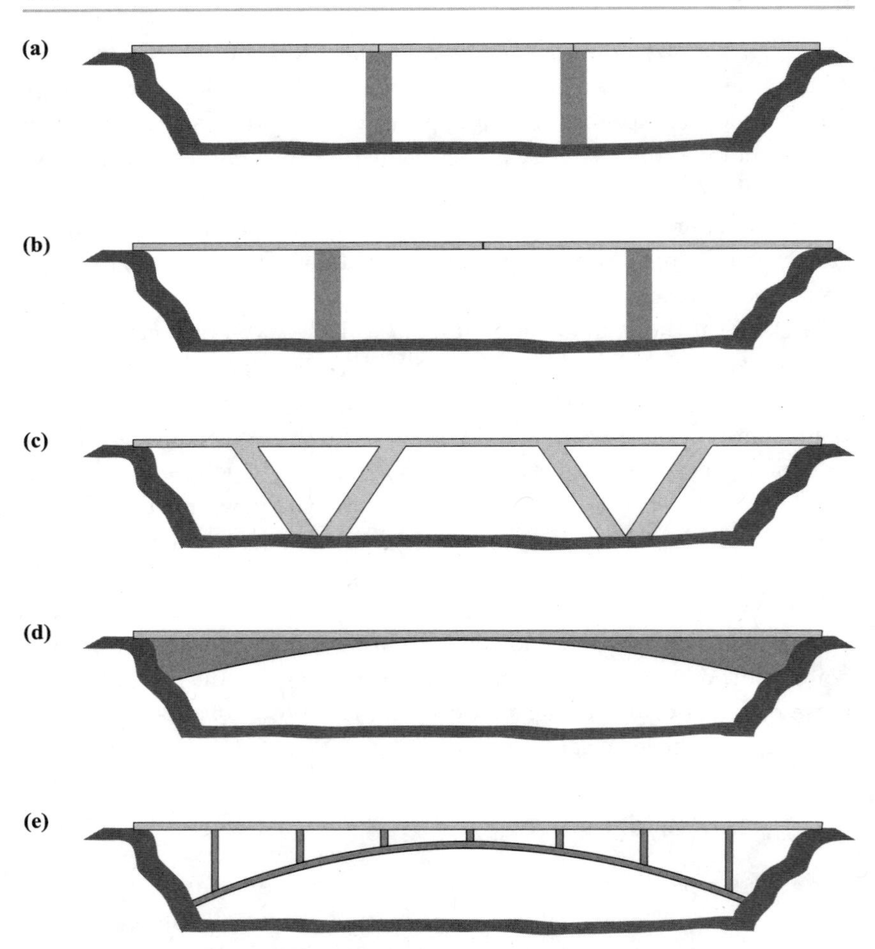

FIGURE 5.1. Flat bridges, with the deck on top: (a) simple beam; (b) cantilever; (c) rigid frame with V-leg; (d) closed spandrel deck arch; (e) open spandrel deck arch.

Underneath the Arches

Arches have been made of stone for three millennia, as we saw in the last chapter. Stone is strong in compression, and this fact restricted the shape that (thin) arches could take. If they were built with the optimum shape for compression—the funicular curve—then they were very strong and stable. We saw a spectacular example of a funicular arch in the Gateway Arch in St. Louis, but this is not a practical shape for a bridge. For an arch bridge to span a great space, it must be very flat, and flat arches require very strong abutments to resist the horizontal thrust that arises from the arch weight. For stone arch bridges a better solution is to string together a series of arches (an arcade with a deck on top, as in fig. 5.4) rather than attempt a single large span. More recently, different arch designs have arisen, and

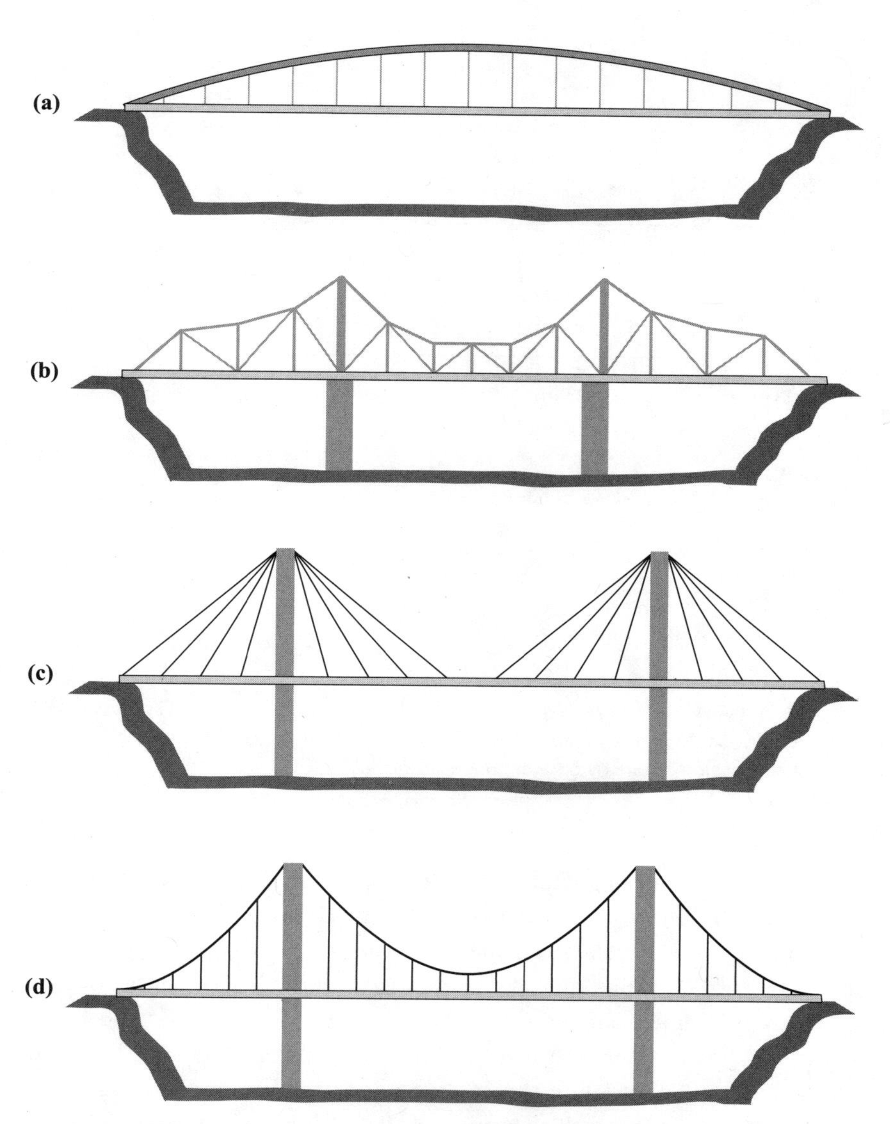

FIGURE 5.2. Through-deck bridges: (a) bowstring or tied arch; (b) cantilever and through truss; (c) cable-stayed; (d) suspension.

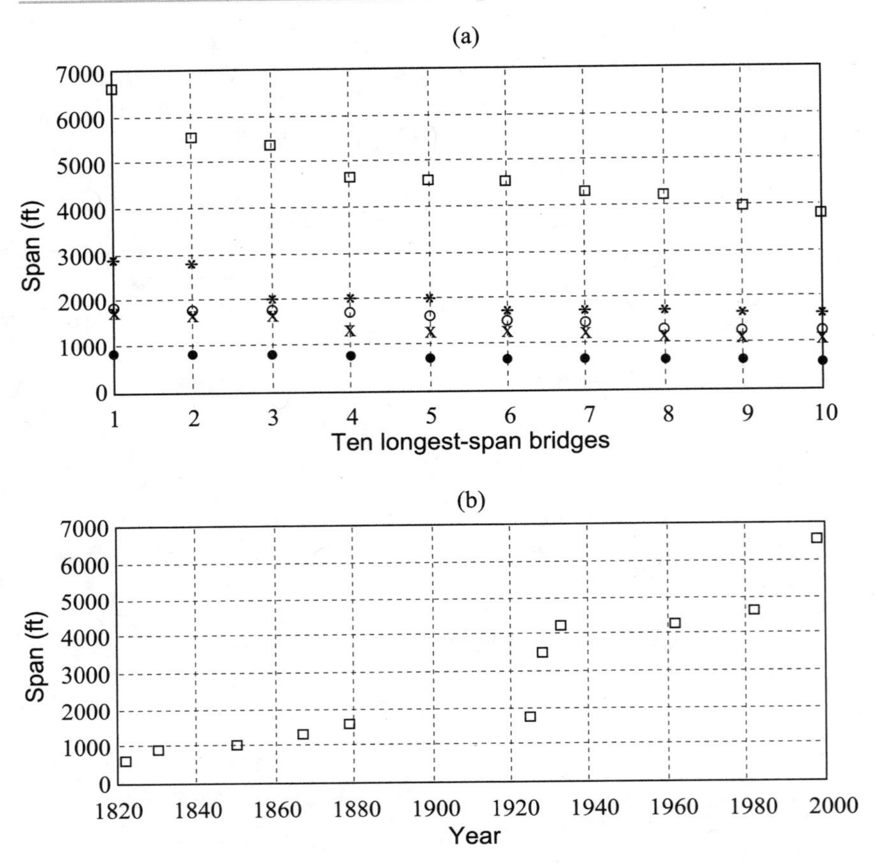

FIGURE 5.3. (a) The ten longest-span bridges in the world today, by type: steel truss (•), arch (x), cantilever (o), cable stayed (*), and suspension (□). (b) Span of the largest suspension bridge by date of construction.

these new designs permit arch bridges to span must larger distances than the traditional stone arches were capable of spanning.

Note the semicircular arch shapes in the bridge of figure 5.4. The Romans built bridges like this, and their design persisted for centuries after the Roman Empire had been relegated to history. Indeed, over the subsequent 1,500 years there was little development in stone arch bridge design. Ribbed arches appeared in the Middle Ages, as we saw earlier. The only other improvement was the adoption of streamlined piers for river bridges. Roman arches were "streamlined" only on the upstream side; the downstream side was left flat or bluff. This feature shows a lack of understanding of fluid dynamics because a sharp cutoff leads to eddies that significantly increase fluid drag. This is why airplane wings taper downstream and why many boat hulls taper at the stern. I don't suppose that the bridge builders of the Middle Ages knew any more fluid dynamics theory than their Roman predecessors; they must have happened upon the improved streamlining that is evident in figure 5.4 by trial and error.

Arch bridge evolution took off in the eighteenth century. Open spandrel arches were adopted for river bridges, making them lighter and also less prone to being swept away by floodwater. An important development for bridge building was the formation of the Ecole Nationale des Ponts et Chaussées in France. This institution was effectively a graduate school for furthering the scientific understanding of bridge and road building. A lot of early theoretical work at the school produced the foundations of our mod-

FIGURE 5.4. The Charles Bridge in Prague, Czech Republic, a traditional stone arch bridge. Note the angled stonework at water level: this is not just to break up ice or to protect the bridge from damage by logs or wayward boats, but to reduce erosion damage by streamlining water flow. (The angled stonework is on the downstream side as well as upstream because, without this construction, eddies would build up, generating hydrodynamic drag and increasing erosion.) Thanks to Glenn Sanchez for this image.

ern understanding of bridge engineering principles. Equally important from a practical bridge-building perspective was the Industrial Revolution in England. The invention of railways led immediately to the building of many new bridges that were strong enough to take heavy locomotives. George and Robert Stephenson, a father and son team best remembered for the development of locomotives and of railway infrastructure in England, built a number of stone arch bridges. One of these was the 28-arch bridge at Berwick in northern England, on the main line between London and Edinburgh. This bridge, completed in 1850, is still going strong today.[2] The great Victorian engineer Isambard Kingdom Brunel, that larger-than-life builder of just about anything,[3] built a very flat, elliptical arch bridge out of bricks over the river Thames at Maidenhead in southern England. Timid souls predicted that it would collapse, but it still stands today. An earlier giant builder of industrial Britain, the Scottish engineer Thomas Telford (known as "the Colossus of Roads"), built many stone arch bridges, such as the elegant Dean Bridge in Edinburgh, Scotland, still in use. He also proposed a very flat 600-foot cast iron arch to replace the old London Bridge. Nervous sponsors scuppered this plan, though modern analysis suggests that it was perfectly feasible.

One of the themes that I have noticed in the development of technology —in several sectors, not just bridge building—is the contrast between continental European theorists and British empiricists. Many of the early builders of the Industrial Revolution in England and Scotland were knowledgeable about the theoretical advances made by, for example, the French academics and engineers at the Ecole Nationale des Ponts et Chaussées, but many were not; they proceeded by their own seat-of-the-pants methods. Both approaches have their place and indeed can feed off each other: it can be argued that the later developments of American suspension bridges

2. I passed over it many times in my student days. It seems to me that bridges are a lot more noticeable from a train than from an automobile. I suppose that this is because we get to see more of the bridge approaches—no traffic in the way—and because we don't have to pay attention to the driving.

3. Brunel was a classic "candle burning twice as bright for half as long," who left behind him a legacy of dockyards and railways, and of innovative tunnels, ships, and bridges that did much to accelerate the Industrial Revolution in England. One of his daring bridge constructions was an early suspension bridge at Clifton in Bristol. High over a gorge and made from wrought iron links (steel had yet to be refined for cables), this bridge was completed in 1859 after Brunel's death, and it still functions.

benefited from both traditions. Ultimately, the more analytical design philosophy dominated, if only because of the high cost of failure when a big bridge collapses. Today, no designer of a bridge, a tower, or a skyscraper would think of proceeding without a thorough scientific analysis (worked out on a computer, of course, because the detailed equations cannot be solved any other way). In the early days of industrialization, however, our theoretical understanding was just being born, and the urgency of the industrialization process gave much more scope to the empirical engineers. Canals were needed, and so aqueducts had to be constructed. Railways spread across the land like wildfire, and so viaducts were built by the hundred. New materials spurred further development: cast iron (strong in compression) could replace stone; wrought iron (strong in tension) permitted new types of bridges to be constructed.

Another important material for arch bridge evolution is concrete. The first use of concrete in bridge construction occurred in France in 1840, and much of the development of concrete arch bridges arose in that part of the world. Robert Maillart was a Swiss engineer, active during the early years of the twentieth century. He is perhaps the most celebrated designer of reinforced concrete bridges and is said to have developed an uncanny understanding of his material. This understanding produced bridges that were both economical (concrete has always been much less expensive than steel) and aesthetically pleasing. A generation later, Frenchman Eugène Freyssinet invented prestressed concrete and was the greatest pioneer in its use for bridge building. Note in my description of these two engineers' work the words *reinforced* and *prestressed*. The practices defined by these adjectives change the nature of concrete, as we have seen. Reinforced concrete and (especially) prestressed concrete are strong in tension as well as compression, and this fact opens up extra possibilities for bridge design.

For 30 years in the mid-twentieth century, the Sandö Bridge, in Sweden, held the record for a concrete arch bridge span, at 880 feet (with a rise of 140 feet—a flat arch). This bridge is an open spandrel deck arch (fig. 5.1e) made of reinforced concrete. During construction in 1943 the centering collapsed, killing a number of workers, and the arch had to be rebuilt, with more significant scaffolding this time. This tragedy highlights a persistent difficulty with building arches: significant temporary scaffolding is always required until the arch is completed. Since 1979 the record span for a concrete arch bridge has stood at 1,280 feet; the bridge itself stands in Zagreb, Croatia. Steel wins out, though, due to its greater versatility. The

FIGURE 5.5. This truss bridge (the Dom Luís Bridge in Porto, Portugal) is both a deck arch bridge (the main deck goes across the top) and a bowstring arch bridge (the lower deck, which looks like a walkway, holds the arch ends together). It is quite common for steel bridges to be of a hybrid structure, like this. In the background is an open spandrel reinforced concrete arch bridge. Thanks to Waldemar J. Poerner for this photo.

longest steel arch bridge span is the Lupu Bridge in Shanghai, China, with a span of 1,804 feet. This bridge wrested the record from the New River Gorge Bridge in West Virginia, a steel truss deck arch.[4]

Steel cable or steel truss beams can stretch across the base of a steel arch to form a *bowstring*, or *tied*, *arch* bridge, shown in figure 5.2a and also in figure 5.5. If the deck of an arch bridge pierces the arch, the resulting structure may also be a bowstring arch. The Lupu Bridge is of this design. The horizontal deck or truss of these bridges holds together the two ends of the arch, relieving the lateral forces upon the abutments. So, bowstring

4. Deck arches illustrate the versatility that steel adds to arch bridge design. The deck can be thinner than a simple beam bridge of the same span because its weight is borne by the arch underneath. On the other hand the arch can be thinner than that of a simple arch bridge because the steel deck stiffens it, so less strength is needed to resist buckling and bending.

FIGURE 5.6. A thin arch bridge in Venice. The curve of this arch must be close to funicular. Thanks to Glenn Sanchez for this image.

arches contain horizontal members in tension, but as a result they require less substantial abutments.[5]

We have seen that a thin masonry arch (such as that shown in fig. 5.6) must adopt a shape that is close to the funicular shape—a catenary—because it can withstand only compressive stress. What about a thick arch,

5. I could mention many specific examples of modern (that is to say, steel or concrete) arch bridges, but here I will restrict myself to three steel arches of very different appearance. All can be found by an online search if you are interested in pursuing arch bridge structures. The Eads Bridge over the Mississippi in St. Louis, completed in 1874, is a string of steel arches that still functions today, carrying more than its design load. The Garabit Viaduct in France was designed by Eiffel, of tower fame, and was completed in 1884. You will not be surprised to learn that it is an open truss structure and is very slender and elegant. It also rises 400 feet above the river Truyère on the Paris-Marseilles railway line. It is so sturdy, despite appearances, that a 400-ton locomotive in the center of the viaduct will deflect the arch only ⅓ inch. My third example is the pride of Australia: the Sydney Harbor Bridge. This is a (very large) bowstring arch, based upon the design of the Hell Gate Bridge in New York and the Tyne Bridge in northern England.

such as the closed spandrel deck arch bridge of figure 5.4? Do the curves here need to be funicular, and what is the shape of the funicular for this type of arch? If the material that fills in the spandrels of these closed arches (the *surcharge* that adds to the bridge's dead weight) is strong in compression, like the voussoir stones, then the arch curve need not be funicular. The lines of force from any part of the bridge must find their way to the ground without passing through the arch, or the bridge will collapse. For a closed-spandrel bridge, however, this does not mean that the lines of force have to pass through the voussoirs (as they must for a thin arch); they may pass through the closed spandrels, and the bridge will still stand. So the designer of closed spandrel masonry arch bridges has more freedom to choose the shape of the arch curve shape.[6]

What about open-spandrel arches, such as that shown in figure 5.7? Let us ignore the fact that this bridge is made of steel so that it has tensile strength and therefore the arches need not follow funicular curves. Let us pretend that the bridge is of unreinforced masonry, thus requiring a funicular arch for stability. Is the funicular shape a catenary for this bridge, as for previous arch bridges that we have considered?[7] In fact, no. In the limiting case where the deck is much more massive than the supporting arch (the bridge of fig. 5.7 may be like this), the equation that describes the funicular curve is that of an inverted *parabola*. (See appendix note 8 for a reminder of the equation for a parabola.) If, on the other hand, the deck mass is much less than the arch mass, the funicular curve will be very close to our old friend the catenary (of appendix note 6) because the dead weight comes mostly from the arch, as in chapter 4. What is the difference between the catenary shape and that of a parabola? Both curves are plotted in figure 5.8; you can see that there is little difference. What if the bridge deck and the bridge arch have comparable masses, so that neither can be neglected? Not surprisingly, the funicular curve is in between a parabola and a catenary.

6. For the record, if you choose to build a closed spandrel masonry arch bridge and you want the lines of force to pass through the voussoirs, for reasons best known to yourself (perhaps the surcharge is a substance with little compressive strength), then the arch curve is still a catenary. However, this curve describes the funicular shape only if the height of surcharge above the arch is just right. The bridge deck (assumed to be atop the surcharge) must be high for flat arches and low—close to the arch apex—for very curved arches.

7. Stay with me: this discussion may seem arcane, unrealistic, and beside the point, but its relevance will soon be clear if you haven't already noticed it. Hint: Robert Hooke would immediately have seen where I am going.

FIGURE 5.7. The High Bridge, Manhattan, an open spandrel deck arch bridge. Thanks to John R. Plate for this image.

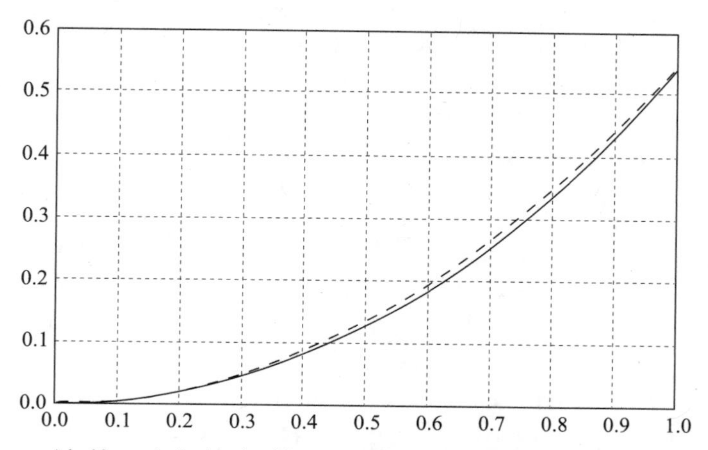

FIGURE 5.8. A half parabola (dashed line) and a half catenary (bold line), with parameters chosen so that both curves start and finish at the same points.

All this geometry just emphasizes the role that mathematics plays in the physics of bridge funicular shapes. However, modern bridge-building materials are strong enough in tension and compression that the bridge arches do not have to be funicular. When we get to suspension bridges, the story will be different.

Trestle and Box Girder Bridges

There are three primary load-carrying systems for bridges. Loads can be borne by arches in compression, as we have seen. Loads can be suspended from cables or stays in tension, as we will see later in this chapter. Here, I will explain the third—and in some ways intermediary—option: loads can be transmitted through beams or trusses that withstand both compressive and tensile stresses. Apart from the short and simple beam bridges and truss bridges of earlier chapters, there are two significant variants that deserve mention here: the *trestle* bridge so characteristic of nineteenth-century America and the box girder bridge of modern Europe.[8]

Trestle bridges consist of a lot of short spans that are supported by splayed vertical members. They are normally constructed for railroad use, as is the bridge shown in figure 5.9. The old wooden bridges thrown up rapidly may have lasted only a short time, but many of the more carefully constructed wooden trestle bridges of the late nineteenth and early twentieth centuries are still with us today. These timber trestles were built from creosoted logs bolted together to form a complex three-dimensional truss structure. If the foundation was firm and the bridge engineers knew what they were doing, such bridges were strong and robust. They easily supported the weight of a locomotive (though we will look at some of the issues that arise from dynamical loading later in this chapter) and were not susceptible to wind loading because of their open scaffold structure. A significant advantage of trestle bridges is that they require no centering: as the bridge is built, each section can support its own weight. As you may imagine, these bridges were inexpensive and relatively quick to build.

Trestle bridges—some of them very long—are still built today, though of steel rather than timber. They are not elegant and will win no awards for architectural style or span length, but they are functional, light, and sturdy.[9] In Western Canada, where I live, many creeks and coulees were

8. Trestle bridges evolved from truss roof structures in the early nineteenth century. The modern box girder bridge evolved by joining the flanges of parallel I-beams.

9. An early exception was the Tay Bridge in Scotland, which collapsed in 1879 from side winds of 70–80 mph, carrying a passenger train with 75 people on board to their deaths. This bridge was constructed of cast iron pillars and wrought iron trusses. An official inquiry concluded that the bridge was "badly designed, badly constructed and badly maintained." No account had been taken by the designers of the combination of wind loading and a moving live load; also, the foundations were

FIGURE 5.9. A steel trestle bridge, this one on the Georgetown Loop Railroad west of Denver, Colorado. Thanks to Marvin Berryman for this image.

spanned by CPR (Canadian Pacific Railroad) engineers with steel trestle bridges in the first decade of the twentieth century. The biggest is the High Level Bridge near Lethbridge, Alberta, which consists of 33 steel truss towers supporting a railroad deck more than a mile long and 300 feet above the valley floor. This bridge was built in 2 years (completed in 1909) and is still used today.

Box girder bridges are related to trestle and other truss bridges, though they look very different (fig. 5.10). Plates replace truss members, so that the box sides are closed, not open. The response of such boxes to different kinds of stress is not the same as that of truss structures of the same overall

---

too shallow. The modern replacement bridge (which is very sturdy) sits beside the broken piers of the old bridge, somewhat alarmingly for the railway passengers, who can see the wreckage from a train on the new bridge. In the 1880s there were an average of 25 iron bridge failures in the United States each year. Steel proved to be a more reliable bridge-building material than iron. More importantly, modern analysis has led to a greater understanding of the stresses that live loads can induce.

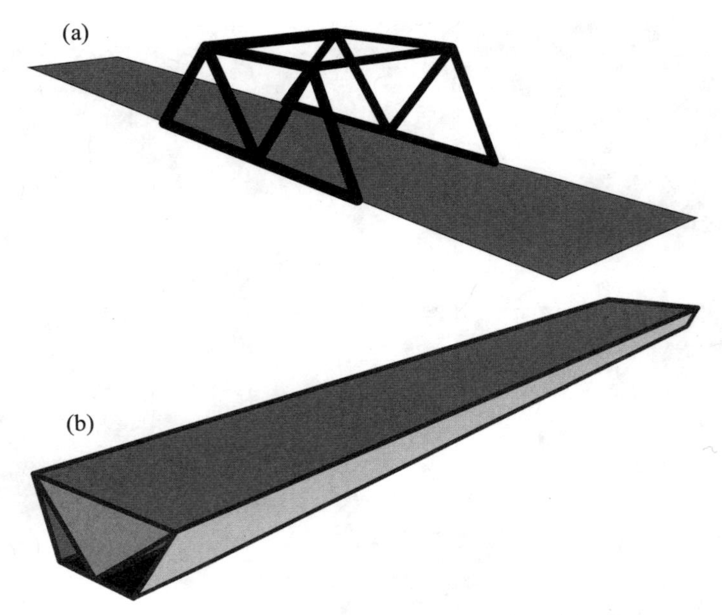

FIGURE 5.10. Box girders, old and new. (a) A simple box girder bridge design constructed from truss members. (b) A modern design for a box girder bridge constructed from steel plates.

shape. The analysis is much more complex, and construction is more difficult, which has led to problems in the past, as we will see. Basically, plates that are welded or bolted to other plates along their edges form a structure that is more rigid than a truss structure. Solid box girders can resist shear forces (see chap. 1) and are particularly rigid against torsion (twisting) stress. We have seen the advantage of I-beams and girders: they resist bending in one direction. Box girders resist bending and twisting in all directions. The shape does not have to be a box—a circular pipe is an effective box-girder. The important feature is that the structure be closed, with most of the steel on the outside, rather than open like an I-beam. The word that engineers use to describe girders that are resistant to multiple stresses acting at the same time and in different directions is *orthotropic*. Orthotropic bridge decks were first built in Germany as a solution to materials shortages following World War II; box girders save on weight and steel for a given span, compared with other bridge designs, because of their stiffness. The resistance of box girders to torsion stress is particularly important for road bridges that are curved in the horizontal plane, as often occurs at complicated road intersections.

We are familiar with box girders in another context. You must have seen cranes with a telescopic arm that extends the crane's reach considerably. These arms obviously need to remain stiff—not bend under a load—and yet they need to be light so that the crane can do useful work, moving a load and not just moving itself. Bridge box girders are stiffened more than crane box girders by including internal flanges that divide the tubular structure into box-shaped compartments, in just the way that bamboo wood is internally divided into sections. This helps to resist buckling, but it also makes the box girders more expensive and more difficult to maintain (because of the confined space).

The first box girder bridge was the Britannia Railway Bridge, built over the Menai Strait in Wales in 1850. It had a main span of 500 feet, and trains traversed it inside the box. The design was revolutionary, though it was reproduced only once (and that was for a bridge in the United States). Until 1940 the structural possibilities for box girder bridges were limited. Structures were assembled from rolled sections and plates that were riveted together. The possibilities increased with the invention of electric welding. We have seen that box girders are capable of greater spans than I-beams; they can also achieve a given span with less depth.[10] The typical span of orthotropic box girder decks is currently 250–800 feet; the record span is 1,000 feet for a bridge in Rio de Janeiro (though this probably stretches the limit of economic viability for box girder design).

Box girder bridge decks, typically rectangular or trapezoidal in cross section (again, classical geometry impinges upon engineering) are used for light rail bridges as well as for highway flyovers, and sometimes for the decks of cable-stayed bridges. They can be made from prestressed concrete as well as structural steel or from a combination of the two materials. Nowadays, they are very common, particularly in Europe, where they originated, and are safe and economical up to a certain span. Their evolution was not without difficulty, however. Around 1970 there were, within a few months, four significant box girder bridge failures during construction (in Austria, Germany, Wales, and the worst of all, in Australia), resulting in a total of 51 deaths. The bridge collapses occurred as the box girder sections

10. Span-to-depth ratios for box girder road bridges are in the range of 20 to 35, and of 15 to 20 for rail bridges with their heavier loads.

were cantilevered out from a supporting pier. That is to say, a supporting pier would be constructed—in the middle of a river, for example—and then box sections of the bridge deck added symmetrically to each side of the pier, heading towards each shore. The deck sections have to be added symmetrically, for balance, so that the pier will not topple over. When completed, the bridge is sturdy, but in these cases as the box sections were being cantilevered out, there arose very strong buckling forces acting upon the sections. The analysis of the stress distributions was not properly understood because of the complexity of these forces for box girder designs, and a number of collapses resulted. We now know that the stresses which arise on the girder sections that are directly over the piers can be several times the design load for the completed bridge.

Cantilever Bridges

Not many cantilever bridges are built these days, but for a generation either side of the year 1900 they held the record for largest spans. At its simplest, a cantilever bridge is a construction (usually a truss) that spans a space by taking advantage of the lever principle, like a crane with a counterweight. Usually, cantilever bridges are built symmetrically so that the load acting upon supporting piers is always vertical, even during construction, as sketched in figure 5.11. This characteristic of cantilever bridges is one of their advantages: the supports are simple piers that have to resist only vertical forces—there are no horizontal stresses upon them. Simple vertical loads also mean that it is relatively easy for the bridge designer to take into account thermal expansion and ground movement. Another advantage accrues from the method of construction: the space being spanned (a highway or river) is not disturbed very much by the bridge construction, because apart from the piers all the construction takes place up in the air. The main disadvantage of cantilever bridges is that they tend to be massive and therefore expensive.

The best-known cantilever bridge is the Forth Rail Bridge, over the Firth of Forth just north of Edinburgh, Scotland (fig. 5.12), which was completed in 1890. Like its contemporary, the Eiffel Tower, this bridge is constructed of trusses held together by rivets (8 million of them) and was a prestige project that awed the world. The Forth Rail Bridge grabbed the attention of the Victorian world not just because it achieved the largest span of any bridge up till then but also because it did so twice: the two main spans are

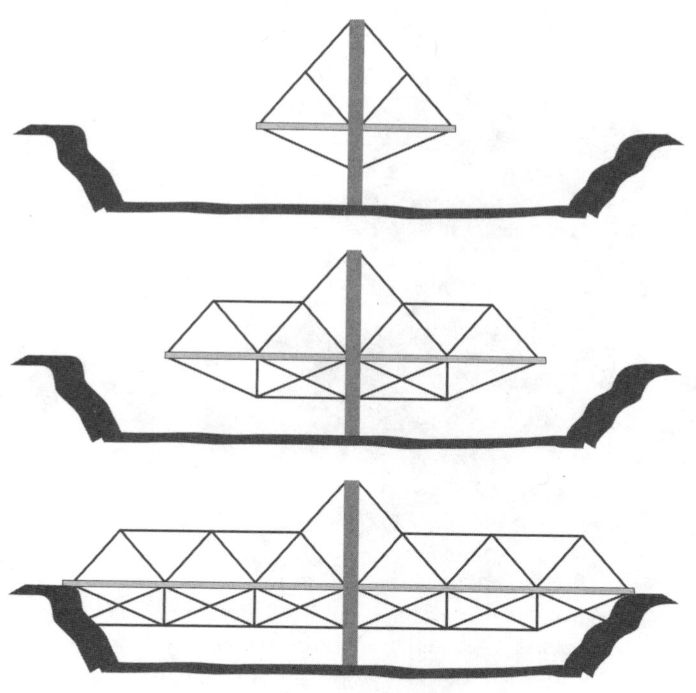

FIGURE 5.11. Stages in the construction of a cantilever bridge.

each 1,710 feet. The 58,000 tons of steel[11] that went into this behemoth were constructed as a giant balancing act; the bridge was built out from each pier, as in fig. 5.11, and then joined up mid-air, 150 feet above the water. At the time of its completion this bridge was considered by many to be the eighth wonder of the world (a feeling shared by many subsequent bridge builders about their own creations, and not without justification).

Some people criticized the Forth Rail Bridge for being over-engineered. It was just *too* strong; a satisfactory bridge could therefore have been built less expensively. It is massive and is, even today, one of the strongest bridges in the world. It is valid to point out that bridges (or skyscrapers or towers) should not be over-engineered. *Of course* safety is an important issue (the Tay Bridge disaster was on every Scotsman's mind at this time—the designer of that bridge put forward plans for the Forth bridge, hastily scrapped), but it is all too easy to spend somebody else's money

11. All the largest cantilever bridges are made of steel. Many medium-sized cantilever bridges are concrete.

FIGURE 5.12. The Forth Rail Bridge, Edinburgh, Scotland. A marvel of Victorian engineering, this massive steel cantilever bridge is famous for its maintenance: it always requires painting. (The pale sections in the middle of the bridge are not concrete piers, but tarps to shield maintenance workers from the sharp Scottish weather.) Scaffolding is everywhere. Note also, in the left background, the Forth Road Bridge, a suspension bridge built 80 years after the Rail Bridge. What a contrast of styles! Thanks to Alan McFadzean for this image.

over-designing a structure to higher-than-required safety standards. On the other hand, in the days before structural analysis and computer number-crunching provided engineers with detailed knowledge about the stresses that act at all points within a bridge structure under all conceivable combinations of loads,[12] it took a brave engineer to criticize a bridge for being too strong.

Brave and lauded if successful (as in the case of the well-remembered engineers Telford and Brunel), but brave and foolish if unsuccessful. A bridge designer called Cooper was one of those engineers who criticized the Forth Rail Bridge for being too strong. Mr. Cooper went on to design a

12. Well, most conceivable loads. No skyscraper designer in the 1960s would have felt the need to include an analysis of the stress resulting from an intentional plane strike, for example. More in chapter 7 on this sad reflection of modern-day building requirements.

cantilever bridge over the St. Lawrence River in Quebec, Canada, which was built in 1907. It collapsed with loss of life during construction because it was built too economically—not enough steel. You can see from figure 5.11 (now that you are experts in truss analysis) that during construction the lower chord of a cantilever bridge will be in compression and the upper chord in tension. These stresses will be very large, especially just before the cantilevered sections are completed, and buckling is a risk. The replacement bridge at Quebec (also of the cantilever type, though built much more sturdily) was completed 8 years later, still stands, and holds the record for the longest span of any cantilever bridge.

Cable-Stayed Bridges

Chronologically, I should discuss cable-stayed bridges after suspension bridges, but in terms of increasing spans, cable-stayed bridges fit in here. Also, cable-stayed bridges have much in common with cantilever bridges, so it makes sense to turn to them at this point. A cable-stayed bridge (depicted in fig. 5.2c) is a cantilever bridge with cables added to relieve the load. The same principle is utilized for cranes that you see on construction sites (see fig. 5.13a). For such a crane, the load (a wrecking ball in fig. 5.13a) is balanced by a counterweight. The crane arm stands upon a slender vertical pier that experiences only vertical loads because of the balancing. So the crane's reach is achieved by the cantilever principle. Stays added to the top of the crane, as shown, reduce the bending force upon the crane arm.[13]

Usually the cables of a cable-stayed bridge are distributed symmetrically about each tower, so that the weight of deck supported by a tower is symmetrical. As with cantilever bridges, the tower experiences no horizontal loading, a great advantage.[14] The deck is in compression and is drawn towards the tower by the cables. So, the deck needs to be stiff and resistant to buckling. These bridges are constructed in the same way as

13. Another common device that illustrates the difference between cable-stayed and suspension bridges is the household rotary clothes line. It acts like a suspension bridge when clothes are hung on the chords but like a cable-stayed bridge if they are hung on the struts.

14. There is some asymmetrical loading due to the bridge traffic. For example, heavy automobile traffic along one lane of a road bridge during the morning, and along the opposite lane during the evening, exerts a torsion upon the bridge deck. Such asymmetrical stresses can produce bending in the relatively flexible decks of suspension bridges, as we will see, but not in cable-stayed bridges.

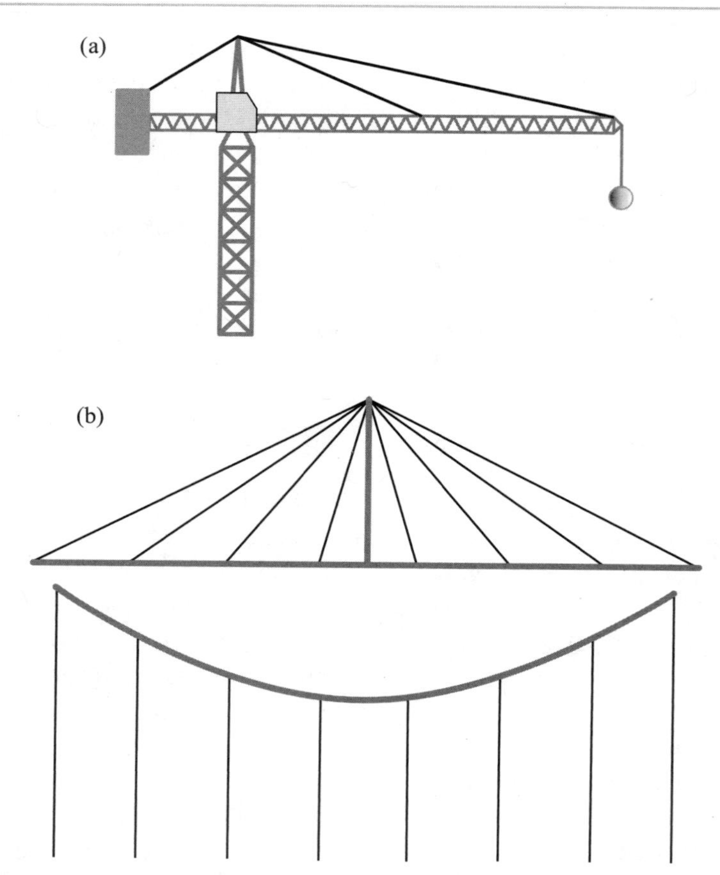

FIGURE 5.13. Cable stays. (a) This crane is a cantilever (if it loses the black stays): the rectangular concrete block acts as a counterweight to the steel ball, and the structure is in balance. The cable stays are in tension, and support the horizontal truss. Cable-stayed bridges follow the same principle. (b) Odd mathematical fact about cable stays. Cut the cables at the top of this bridge so that they are free to dangle down from the deck. Lose the vertical pier. Bend the deck into a catenary and lo! the cable ends lie on a horizontal line.

cantilever bridges: each tower is built independently, and the deck is cantilevered out symmetrically on both sides. Any number of towers may thus contribute to the bridge. Care must be taken during construction, as with cantilever bridges, because the forces that are exerted upon the deck can be considerable.

Cable-stayed bridges are difficult for structural engineers to analyze: because both the bridge and the tower are stiff, they are statically indeterminate. Cable-stayed bridges are not like suspension bridges, where the cable automatically takes up the weight of one section of the bridge deck. For this reason, determining the tension required for each cable of a cable-stayed bridge is not easy, and it is not easy to maintain the correct tension once it has been determined. A slight slackening of the cable—due, for example, to thermal expansion or foundation settling—may require an adjustment of the cable tension. (This is not the case for suspension bridges, as we will see; they simply adjust their shape if loads vary, or if foundations shift or cables expand in the heat.) In fact, the cables that are used in modern cable-stayed bridges are not standard suspension bridge cables; they are of a special locked coil construction that does not stretch much. The cables need to be carefully treated to protect them from environmental corrosion.

Another advantage that cable-stayed bridges share with cantilevers is that they do not require massive abutments or terminal piers because the horizontal loading is very small. Cable-stayed bridges may not be quite as stiff as cantilevered bridges, but they are stiffer than suspension bridges. They may not be as light as suspension bridges, but they are a lot lighter than cantilevered bridges of the same span. They do not fall between these other types of bridges in every category, however: the complexity of the structural analysis for cable-stayed bridges (compared with that for cantilever and suspension bridges) means that these bridges have not proved popular historically. Only in recent decades have engineers developed the understanding, the supporting technology, and the confidence to build cable-stayed bridges. Much of the development of this complex bridge type came from Europe, in particular from Germany. But one early forerunner that used wrought iron chains as cable stays was the Albert Bridge in London, a hybrid design dating from 1872. Another hybrid bridge is the Brooklyn Bridge in New York, of which more in the next section. You can see in figure 5.14 that the approaches to the main span of this bridge are cable-stayed.

Cable-stayed bridges are gradually gaining in popularity and may be encroaching upon the current favorite, the suspension bridge. Thus, a cable-stayed design may be considered for a particular bridge location that, 50 years ago, would automatically have been considered suspension bridge territory. Cable-stayed bridges can be elegant, in a way that cantilever bridges just cannot: they have a slender look with graceful lines. The cables may fan out from the tops of the towers or they may emerge parallel to each other, harp fashion, from different points up the towers. The current longest span for a cable-stayed bridge is that of the Tatara Bridge in Hiroshima,

FIGURE 5.14. The Brooklyn Bridge, New York. Often considered to be a suspension bridge, it is in fact a hybrid design: the approaches are cable stayed. Thanks to John R. Plate for this image.

Japan, at 2,919 feet. This is over half a mile—very impressive even if it does not yet approach the amazing spans of the largest suspension bridges, our next topic.

Suspension Bridges

In 1883 the magazine *Harper's Weekly* announced this sentiment to its readers: "The work that is most likely to become our most durable monument, and to convey some knowledge of us to the most remote posterity, is a work of bare utility; not a shrine, not a fortress, not a palace, but a bridge." The bridge that the good folk at *Harper's* praised to the rooftops (perhaps that should be "tower tops") was the Brooklyn Bridge over New York's East River, completed in that year. As admired in America (it is a National Historic Landmark) as the Forth Rail Bridge was eulogized in England and Scotland, the Brooklyn Bridge, shown in figure 5.14, was the brainchild of John A. Roebling. I have now, in my roughly chronological story of bridge engineering, reached the age of the bridge engineer as well as the age of the suspension bridge. I can only suppose that it is the massive spans that were achieved by these bridges, built sporadically from the early nineteenth century but reaching a climax of construction during the Depression years, that endeared to the public not only the bridges themselves but also the engineers that built them. Hence, we know more about those energetic and courageous men who gave us (actually, sold us) these engineering giants than we do about most other engineers.

I refer you to the bibliography for the stories of these men and their extraordinary efforts and achievements. One more quote will give you a flavor for the high esteem accorded these bridge engineers, and it refers to Roebling and this, the first of the major suspension bridges. In his *Book of Bridges*, Martin Hayden describes the construction of the Brooklyn Bridge as "one of the greatest stories in bridge building and the greatest triumph of nineteenth century American engineering. The building of its 1,600 foot span, suspended from steel wire cables, was a continuous battle from 1869–83 against personal tragedy, natural disasters, unforeseen technical problems and graft." Roebling died before this great bridge was completed, but his company, headed by his son Washington Roebling, saw the project through to completion.

So, what was and is special about the Brooklyn Bridge in particular, and why was the half century after its completion the boom era for suspension bridge building? The answer is a combination of advances in theoretical understanding and in technology. One feature of suspension bridges that has haunted their design, construction, and maintenance from Roebling's time until the present day is bridge deck flexibility. We will see again and again that suspension bridge decks are inclined to tilt (forgive the pun); they may start out flat, but a heavy load or—particularly important—a strong wind can change things significantly. As a consequence, almost from the beginning of suspension bridge building, it has proved necessary in most cases to stiffen the deck in some way—usually by adding trusses alongside or beneath the deck. Such stiffening complicates the analysis and turns what is, at its simplest, a statically determinate problem into a highly indeterminate problem.

We have seen that an understanding of an indeterminate structure requires knowledge of the internal characteristics of each of its main members—stiffness, bending moment, etc. The problem is that the distribution of stresses that arise in indeterminate structures depends upon the deformation of individual members (a significant complication that truss analysis avoids by assuming perfectly stiff members). As early as 1823 the French engineer and physicist Henri Navier (a student and later a teacher at the Ecole Nationale des Ponts et Chaussées) had understood how unstiffened suspension bridges work, and 35 years later an approximate analysis of truss-stiffened suspension bridges was available to engineers through the work of the Scottish physicist and engineer William Rankine.

In the earliest suspension bridges wrought iron eye-bars were strung between masonry towers. The arrival of steel changed everything. High-tensile cables were stronger and more reliable than wrought iron chains, and steel could replace masonry in the bridge towers. But wind loading dogged these early bridges—a signal of the future. Several early suspension bridges in Britain failed due to wind-induced motion. By the time of the Brooklyn Bridge construction, both theoretical analysis and technology were advanced enough for Roebling to attempt another world-record suspension span.[15] In particular, Roebling was aware of the theoretical advances made in Europe (extended to stayed suspension bridges and to purely cable-stayed bridges in 1880 by Clericetti) and knew about the widely admired bridges designed by Telford.[16] He was aware of both the analytical and empirical modes of progression, and his own journey of understanding included both methods. A semi-empirical approach led him to successfully develop his own hybrid style of suspension bridge, and by the time he built the Brooklyn Bridge, Roebling was the preeminent suspension bridge designer in the United States, with several large and important bridges to his credit.

The Brooklyn Bridge has masonry towers and a suspended deck with a main span of 1,595 feet and a total length of 6,016 feet. The approaches to

15. Roebling had earlier attained the record for suspension bridges with his Cincinnati Bridge of 1867, which had a main span of 1,057 feet.

16. Telford designed the Menai Bridge, in northern Wales, as early as 1826. With a span of 580 feet, it was one of the first and certainly the longest suspension bridge in the world.

the main span are cable-stayed, as mentioned earlier. The bridge deck is stiffened by an open truss structure, which is the best way to reduce wind loading effects. This fact was not tested experimentally—wind tunnel tests of bridges began only in the 1950s—and so the aerodynamics were not in any way worked out in advance. Roebling had his own method for estimating stresses and bridge strength requirements, and his hybrid design (the combination of stays and suspension is typical) proved stiff and strong.[17] Roebling allowed a safety factor of 6 for his bridges—tacit acknowledgment of the incomplete understanding of those days.

The Brooklyn Bridge span was surpassed by the main spans of the cantilevered Forth Rail Bridge, which was being constructed at the same time and was completed within a year of the Roebling's culminating achievement. Cantilever bridges would retain the record span until 1929. Within a decade of that economically turbulent year, however, the record for length of span would be regained, and then beaten again and again, by a series of monumental suspension bridges constructed in the United States.[18] We saw the numbers in the graph of figure 5.3b. The climax of this bridge-building frenzy occurred across the continent from New York. What John Roebling was to the Brooklyn Bridge, Joseph Strauss was to the Golden Gate Bridge in San Francisco (fig. 5.15). The diminutive Strauss lived long enough to see his giant project completed in 1937, but only by a year. It is likely that the stressful nature of this mammoth enterprise contributed to his death; the project lasted more than two decades from conception to completion, and required overcoming political and financial problems, as well as technical ones.

The great scale of the Golden Gate Bridge brought unprecedented logistical problems for Strauss, the chief engineer. Numbers become hard to grasp; perhaps stating that 50,000 tons of cement were used for the piers and anchorages (which needed to be massive because of the huge lateral forces transmitted through the cables) means less than saying that this was enough material for a 5-foot-wide sidewalk all the way from San Francisco

17. Stiff, strong, and durable. Though the Brooklyn Bridge required emergency redecking in 1999, it was still, at 110 years, worth renovating rather than replacing.

18. Large suspension bridges of this period include the Ben Franklin Bridge over the Delaware in Philadelphia (1926), the Ambassador Bridge in Detroit (1929), the George Washington Bridge in New York (1931) with a span of 3,500 feet, and of course the Golden Gate Bridge across San Francisco Bay (1937). This last, with a main span of 4,200 feet, held that record for 30 years and held until very recently the record for tallest towers, at 746 feet.

FIGURE 5.15. The Golden Gate Bridge, San Francisco. Image from Wikipedia (Rich Niewiroski, Jr.).

to New York. The span between the two great towers (4,200 feet) was covered by two great steel cables each with a diameter of 36½ inches, between them constituting 80,000 miles of wire.[19] The cables were so long that, in Strauss's words, they would "lengthen on hot days and lower the bridge by as much as 16 feet."[20] Apart from the 389,000 cubic yards of concrete, the bridge consisted of 44,000 tons of steel: the scale of this construction was, at the time, unprecedented in the history of civil engineering.

Once the giant cables had been strung between the giant towers, the deck was cantilevered out from each tower, suspended from the cables as it grew. Here is an advantage of suspension bridge design over cantilever and cable-stayed bridges. During construction the weight of the deck does not induce huge bending forces, because the partially completed deck is sup-

19. The cable for the Golden Gate Bridge was spun by the experts in suspension bridge cables: J.A. Roebling and Sons.

20. At the 50th birthday celebration for the bridge, in 1987, as many as 300,000 people crowded onto the main span. The weight of this very live load was enough to drop the deck height by 11 feet. Suspension bridges are not stiff—far from it.

ported by the cables. The towers of a suspension bridge are purely in compression, and so their design is simple; the cables are purely in tension, and so they can be relatively light. One disadvantage of suspension bridge design we already know about: the cables must be tied down to the ground with massive anchorages that are able to resist being pulled out of the ground by the extremely large tension forces in the cables. Another disadvantage is that the two towers must be carefully located and the cables strung between them before the deck construction can begin. The third and worst problem, less well understood until the second half of the twentieth century, is the effect of wind loading upon the very flexible deck. I will discuss the problems of wind loading in the next section. These issues were known to Strauss and Roebling and to earlier engineers in Europe, who appreciated that the floppy decks of suspension bridges were somehow particularly susceptible to being brought down by the wind. Their solution was to stiffen the deck. The floppiness of suspension bridge decks means that, in general, they are not suitable for heavy live loads such as locomotives. Look around you: how many train bridges do you know that are of the suspension type?

After nearly three decades the world-record bridge span moved from San Francisco back to New York. With its completion in 1964 the Verrazano-Narrows Bridge, rising between Staten Island and Brooklyn, took the title from the Golden Gate Bridge with a central span of 4,260 feet. Even though the bridge is a double-decker, it looks slender because of the great span. This slender elegance is a characteristic of many of the bridges designed by Othmar Ammann. The Verrazano-Narrows Bridge was Ammann's crowning achievement (he had already built four major steel bridges in New York, including the George Washington Bridge). The trademark features of his bridges—slender decks, unornamented towers—were based upon a minimalist approach. He believed that inexpensive solutions to problems were in all respects the best, if they were thought through and well designed. The famous French architect Le Corbusier praised Ammann's George Washington Bridge of 1931 as "the most beautiful bridge in the world." Inexpensive? Well, relatively. Slender suspension bridge decks? Well, as much as possible—with careful design. Ammann designed carefully[21] and had been praised for the aesthetically pleasing slender yet stiff deck of the George

21. An example of Ammann's careful design: the two towers of the Verrazano-Narrows Bridge are 1⅝ inches further apart at the top than at the bottom, to allow for the earth's curvature.

Washington Bridge. He regarded the truss stiffening of many bridge decks to be ugly, criticizing in particular the Williamsburg Bridge over the East River. Ammann's solution for the George Washington Bridge worked even though, at the time, engineers were unaware of the full effects of dynamical wind loading. He considered that the dead weight of the bridge was enough to counter the maximum force that a steady wind—even of hurricane strength—could throw at it. He was right, in this case. His Bronx-Whitestone Bridge was also praised for its very slender deck—but here Ammann had stepped over the line, and the bridge eventually required stiffening with trusses.

Earlier, Strauss said that building a bridge was a constant war against the forces of nature. Nature drew the line at the new generation of slender suspension bridge decks. The collapse of the Tacoma Narrows Bridge in Washington State in 1940 (described in the next section) was a disaster that all bridge engineers ever since have studied. We now know that steady winds and the constant wind loading of a suspension bridge deck are not the problem. Winds can cause dynamic effects with such flexible decks as were to be found at Tacoma Narrows and half a dozen other suspension bridges of the period, including Bronx-Whitestone. The deck can be induced by the wind to oscillate, and these oscillations can grow in amplitude to a degree that eventually causes the deck to tear itself to pieces, as we will see.

No such problem for Verrazano-Narrows, however, because by the early 1960s engineers had enough understanding of aerodynamics—thanks largely to the inquiry set up after Tacoma Narrows—and enough computer power to be able to include the results of detailed aerodynamic studies in their bridge deck design. The Verrazano-Narrows Bridge was the first suspension bridge to be designed with the aid of computers, and the result is a suspension bridge that is slender and yet stable against the effect of wind forces. Stability is achieved both by stiffening (in a way that does not spoil the elegance) and by taking deck aerodynamics properly into account.[22]

Since the 1980s, the record for longest bridge span has passed out of the United States. As with skyscrapers, the burgeoning economies of Asia

22. Another large suspension bridge built in the early 1960s was the Forth Road Bridge, in Scotland. The designers of this bridge were also well aware of the dangers of wind-induced oscillation, and they thoroughly tested their aerodynamic bridge deck in a wind tunnel before finalizing the design. The two Firth of Forth bridges—rail and road—sit side by side and are a powerful visual testament to the evolution of bridge engineering over the 75 years that separates them.

have permitted engineers to construct on a massive scale. The longest span today is fully half again as long as that of Verrazano-Narrows: the Akashi-Kaikyo Bridge (1998), connecting Honshu with the island of Awaji, in Japan, has a span of 6,532 feet—that's 1¼ miles. (In appendix note 9 you will find a quick, back-of-the-envelope estimate of the maximum realistic length of a suspension bridge span.) Ultramodern engineering practices included detailed aerodynamic wind tunnel tests on structural components, special high-tensile steel wire within the cables, and earthquake and tornado studies,[23] all accompanied, of course, by gigabytes of computer number-crunching. The bridge has massive pendulums to dynamically dampen any oscillations that arise near the bridge's main resonance frequency (more on bridge resonance in the next section). To me the most telling statistic signifying modern design is the number of construction workers who die of accidents during a bridge's construction. The Forth Rail Bridge, finished in 1890, saw the deaths of 60–80 workers. Numbers since then have declined, as safety standards and construction practices improved over the decades. Nobody died building the Akashi-Kaikyo Bridge.

One of the arch concepts we have encountered in this book is that of the funicular curve—the shape an arch must adapt if it is to have purely compressive stresses. We saw that the shape for a thin arch was similar to that of a hanging chain flipped upside down. Therefore, it should be no surprise to you to learn that the shape of suspension bridge cables is the same as that of a deck arch bridge (flipped upside down) for those bridges where the deck is much thicker than the arch, so that arch mass is unimportant in determining funicular shape. We saw that the funicular curve for deck arches was parabolic but that it was not always necessary for a masonry deck arch to be of this shape. Flip the arch upside down and we have a suspension bridge. Suspension bridge cables always follow the funicular shape and so are parabolic under uniform load, for the same reason that hanging chain curves are funicular: they are floppy—they cannot resist any bending stress —and so, if there is a sideways force acting on a section of a cable, that section moves sideways. This is what makes suspension bridges so floppy. If the deck is not stiffened, then it will move with changing loads because

23. One earthquake experiment was unintended. The powerful Kobe earthquake of 1995 occurred nearby. The towers had been constructed by the time of this event, but not the bridge deck. The earthquake moved the two towers further apart by about 3 feet, but otherwise they emerged OK. So, the Akashi-Kaikyo bridge is 3 feet longer than it was originally designed to be.

it certainly won't be rigidly held in place by the supporting cables. So 300,000 people on the Golden Gate Bridge cause the deck to sag in the middle: the cables change shape to form the new funicular appropriate for the live load distribution they experience. Say another earthquake moves the towers of the Akashi-Kaikyo Bridge another 3 feet further apart: the cables take the funicular shape appropriate for the new conditions. You can see that floppiness—flexibility—is not all bad news because it allows a bridge to automatically adapt to some changes. Masonry arch bridges would break up if their piers were suddenly moved three feet further apart.[24] Cable-stayed bridges would need significant adjustments, and cantilever bridges may buckle if an exceptionally large load is dumped in the middle of the deck, but suspension bridges take up the slack, literally.

Bridge Dynamics

Sir Isaac Newton would remind us that when we see a live load pass over a bridge, the bridge reacts to the load even though we may not see much bridge movement. Indeed, for most bridges we hope that there will be no movement. In practice, bridges do move when subjected to loads, even if only a little. In chapter 1 we saw how different construction materials compress or stretch when stressed and then return to normal when the stress is removed—if the stress is within the elastic limit of the building material. Bridges of all types—masonry, steel, reinforced concrete—operate within their elastic limits, assuming they are well designed and not subjected to excessively heavy loads. So, they deform a little when a load passes over and then return to their normal shape afterwards. These movements are very small for most materials and for many bridge designs—the notable exception being suspension bridges, which, because of their structure, can deform considerable under heavy loads. Large deformations are not necessarily bad, but we will see here how they can get out of control. First, though, let me provide a simple model of bridge dynamics which shows how sections of a bridge might deform and then return to normal as a heavy load passes over.

24. There are variants of arch bridges that have some flexibility of movement. Such bridges (of steel or reinforced concrete) have two points of contact with the ground ("two-pin" arches) and perhaps also one other point at the apex ("three-pin" arches) which behave like truss pins. The arch section between two pins is rigid like a truss member, and the arch is stable if one of the pins is moved a little (say due to thermal expansion).

Dynamical Loading

We saw earlier that bridge designers of old did not need to worry about live loads because the dead load (masonry bridge weight) was so much greater than any live load, with the consequence that the live loads were negligible. This situation changed in the nineteenth century, when the railroad boom spread out from England to Europe and the New World. Heavy locomotives passed over bridges—typically wooden or steel trestle structures—that were much lighter than their masonry antecedents. Now the live load weights were significant; they were heavier and moved much faster than the carts and wagons that passed over bridges in earlier ages. So, the bridge engineers found out that they needed to take into account the dynamics of bridges, and not just the statics. That is to say, they needed to analyze the forces that change with time as a load passed over a bridge; it was no longer sufficient just to analyze the distribution of forces that arose from the bridge weight and construction, or from a stationary load atop the bridge.

A simple model will give you an idea—a suggestion—of how a bridge can react dynamically to a heavy load moving over it. I will first introduce the model and then will say why it is inadequate. My model is good enough to get across to you the message of bridge section movement in response to a moving load, but it is not sophisticated enough to convey the complexity of this movement. So, model first and explanatory words afterwards. Consider figure 5.16. Here, I have replaced a realistic steel bridge substructure with a series of springs. A spring models elastic deformation very well for the simple reason that it works precisely on that principle. Hooke's law tells us that strain (or deformation—for example, compression under a weight) is proportional to stress (the weight), and springs work this way; double the weight on a spring, and it will be compressed twice as much. One of the simplifying features of the model is that it reduces the elastic deformations of bridge components to a single type. Here, the bridge deck is pressed down as the train passes over it. In real bridges, different components of the bridge will be compressed different amounts, other components will stretch, others will twist, and some components may undergo all three deformations at once. But you get the idea from figure 5.16: the weight of a passing train will cause the bridge to deform temporarily. Analyzing the physics of this funny bridge is straightforward, and the results are shown in figure 5.17, which plots bridge deck deformation with time (using arbitrary units) as the train passes.

FIGURE 5.16. A simple model of bridge dynamics—too simple, in that it suggests that the effects of a live load upon the bridge structure are felt only locally, beneath the load. In reality the sudden appearance of a heavy load may be felt further afield, just as the effects of a pebble dropped into a pond are carried across the surface by radiating waves. This model does show how bridge shape can be distorted by a heavy load passing over, and analysis indicates how each bridge section shakes and wobbles until it eventually settles down once the load has passed.

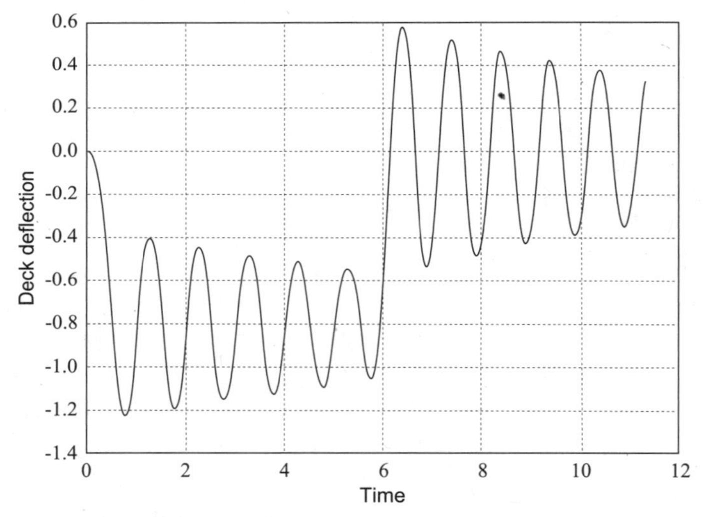

FIGURE 5.17. The deflection (change in height) of a section of bridge deck for the simple spring bridge model of fig. 5.16. As the train passes over, the bridge section is pushed down but bounces back once the train moves on (after 6 time units, in this case). The faster wobbles are due to the spring rebounding; they dampen down due to friction.

In my simple model, the bridge deck sections wobble, and these wobbles then die away as frictions damps down the movement once the train has passed. The different bridge sections (different springs in fig. 5.16) do the same thing but shifted in time; the second section begins oscillating after the first, the third section after the second, and so on, like the staggered chorus of a barbershop quartet. So we get a hint of the complexity of the whole bridge movement: the train influences the bridge movement sequentially so that different parts of the bridge move in different ways at different times. Now multiply all the movements (here, only deck compression is modeled) to include tension, bending, shearing, and torsion, alone or in combination, of every bridge superstructure component, and you can see that a real bridge responds in a complex way to a dynamic load.

So far, so good. Now for the shortcomings of my model. In figure 5.16 only the section of bridge that is directly under the train is influenced by the train's weight. The effect on a real bridge can spread far ahead of the train. Like shock waves spreading through an exploding building, or cracks spreading across glass, the sudden impact of a train's weight at one end of a bridge can spread rapidly, by many different routes through the bridge structure. Now, it is possible that some components of the bridge structure might receive several of these "shocks"[25] from different directions at the same time, and so the effects of the shocks (the deformations shown in fig. 5.17) will add up. Such components will wobble more than other bridge components. If the bridge has been badly designed, these waves may indeed concentrate in certain locations—a particular truss member or joint —and cause it to rattle much more than its neighbors. This kind of behavior is very difficult to analyze in complicated structures, and the means of analysis (computer number-crunching) was lacking in the second half of the nineteenth century when bridge dynamics became important. So, bridge engineers strengthened their structures beyond what they figured they would need for a static load, to ensure that dynamical effects would not cause problems.

25. The term *shock wave* has a technical meaning in physics that is not applicable here, but I can think of no other everyday term to convey my meaning. "Wobble waves" doesn't quite do it. The best analogy I can provide is to think of a pebble dropped in a pool. The waves from the pebble spread out in all directions and reflect off the side of the pool. These various waves interfere with each other, sometimes becoming quite large at certain locations in the pond. These locations of interference correspond to the bridge components that rattle the most.

Suspension Bridge Resonance

But strengthening precautions do not always work for such inherently wobbly structures as suspension bridges. Wind loading of suspension bridges causes them to oscillate, as for the train over the trestle bridge, but because suspension bridges are so flexible, the oscillations sometimes do not die away. Instead, they can grow and grow. At this point I need to clarify the idea of *resonance*, which we first saw in chapter 3 and note 5 of the appendix. To a physicist, resonance is a phenomenon whereby oscillations can become very large—much larger than you might expect—at a particular frequency. The phenomenon is a common one, found in pendulums, organ pipes, tuning forks, spin driers, and many other physical systems. To an engineer, resonance is bad news. In a spin drier it causes violent vibrations at certain spin rates; in a badly designed suspension bridge it can cause catastrophic collapse. The basic physics can be conveyed by adapting our simple spring bridge model as in figure 5.18.

The first car that passes over our hypothetical spring bridge causes the bridge deck to drop under the vehicle's weight and then bounce back when the car has passed. The deck will oscillate up and down, as we saw in figure 5.17. A second car happens along at just the right time to reinforce the downward movement, which gets bigger. A third car does the same, and as more and more cars pass over the bridge, the deck oscillations grow. It is like pushing a child on a swing each time she reaches the top of her backswing; the little push you provide causes the swing amplitude to increase. Of course, timing of your push is crucial to making the swing do its thing; likewise, the interval between cars passing over the bridge must be regular. The phenomenon of oscillation amplitude growth under the application of periodic (regularly repeated) force is what we physicists call resonance. For the car/bridge example, we can plot the effects of resonance by computer calculations as shown in figure 5.19.

Again, my spring bridge model is a grossly oversimplified version of reality, but I hope that it conveys the essence of resonance. What we have here is an example of mechanical resonance.[26] One instance of mechanical resonance of a nineteenth-century bridge in England is almost as famous among bridge engineers as the collapse of the Tacoma Narrows Bridge in

26. There are other types of resonance in physics, including acoustical and electrical resonance. Mechanical resonance is—duh—caused by a periodic mechanical force.

FIGURE 5.18. Resonant oscillations can be displayed with this simple model of bridge dynamics. The bridge is like a spring, which is compressed when a car passes over but which rebounds once the car has passed. The next car causes another compression, which might reinforce the oscillation if it is timed right. A line of such cars, equally spaced and moving at the same speed, will periodically reinforce the oscillation, like pumping a child's swing, causing the oscillation amplitude to grow.

the United States. In 1831 a troop of 60 soldiers were marching over Broughton Bridge in Manchester, England, when it collapsed underneath them. The bridge was an early type of suspension bridge with chain cables. Investigations by engineers after the event showed that a badly manufactured iron coupling bolt had sheared under the load, but interestingly, the strength of this bolt was certainly sufficient to resist the stress imposed by the weight of 60 soldiers if they had been standing still. The investigators concluded that the marching cadence of the soldiers induced mechanical resonance of the bridge, which oscillated more and more wildly until the bolt broke (see Drewry, in the bibliography, for details). Since the Broughton Bridge collapse, soldiers have been ordered to break step when marching over bridges, to avoid setting up resonant oscillations.[27]

So, mechanical resonance can be caused by cars passing over a bridge periodically (according to my simple model), and by soldiers marching. The key point is that, to cause large oscillations, the periodic force of the live load must be close to a natural frequency of the bridge. (My spring bridge has only one natural frequency, but real bridges can have several.) Earthquakes can cause resonant oscillations in tall buildings, and modern skyscrapers may have active dampers to deter oscillations at the building's

27. A more recent example, also from England, is the resonance that occurred on the London Millennium Bridge (a footbridge across the river Thames) shortly after it opened in June 2000. Thousands of pedestrians caused the bridge to oscillate sideways, and their footsteps caused resonance which increased the oscillation amplitude. Fortunately, this bridge was strong and did not collapse. It has subsequently been altered to eliminate these unusual lateral oscillations.

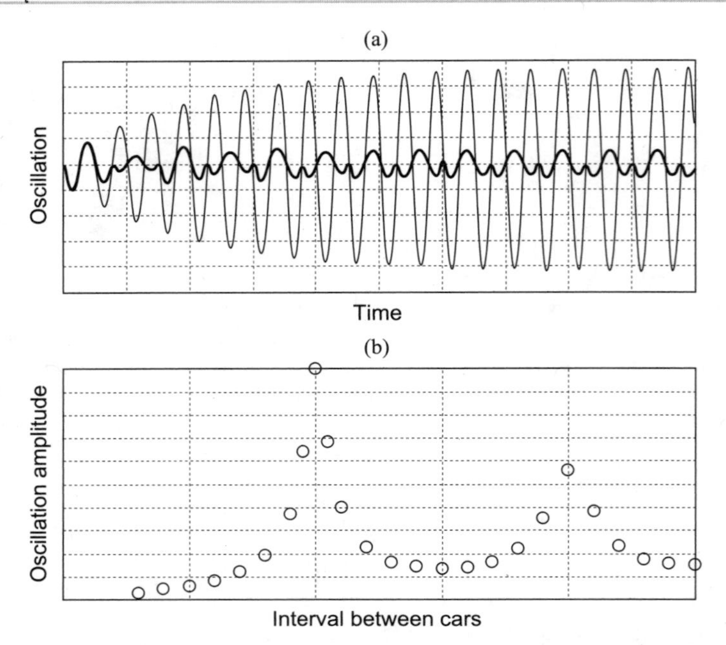

FIGURE 5.19. Resonance of the spring bridge shown in fig. 5.18. This bridge has a natural frequency of oscillation of one cycle every 10 seconds or every 10 milliseconds—the units of time are arbitrary for this example. ("Natural" frequency means the frequency at which the bridge would wobble without any external interference except for an initial shove.) In this case, the cars repeatedly interfere with this motion. (a) Bridge oscillations for cases in which a car passes every 15 seconds (bold line) and every 10 seconds (thin line). The oscillation is much bigger when the external force (the line of cars) has the same frequency as the bridge natural frequency. (b) This point is emphasized by plotting oscillation amplitude against the interval between cars. The main resonance peak occurs for intervals of10 seconds, but secondary peaks occur at multiples of this value.

natural frequency, as we saw earlier. In 1981 a tragedy occurred in a Kansas City hotel when an elevated walkway collapsed under the load imposed by many dancers. It is quite likely that the rhythmical motion of the dancers induced resonant oscillation that led to the collapse.

For suspension bridges, the main external source of forces that give rise to resonance is not cars or people but wind. The Tacoma Narrows Bridge provided dramatic evidence of the effects of wind-induced oscillations just as soon as it was completed, in 1940. At the time, this bridge was the third longest suspension bridge in the world. It spanned Puget Sound in Washington State and was frequently subjected to mild to moderate crosswinds.

From the get-go it was clear that the deck flexed and wobbled under the influence of these winds. The deck would wobble the way that a string can often be seen to vibrate: along its length, with a node (a point with zero oscillation amplitude) in the center. So familiar were these wobbles that the bridge was nicknamed "Galloping Gertie." On November 7, 1940, four months after the bridge opened, a new mode of oscillation caused the deck to shake itself to pieces and fall into the water below. Famously captured on film, the deck can be seen to wobble from side to side (not along its length) with increasing amplitude.[28] An engineering and financial disaster, the deck collapse (the cables remained intact) claimed no human life. People had just enough time to get off the bridge. The only fatality was Tubby, a three-legged cocker spaniel who resisted efforts to rescue him from an abandoned car and who, along with the car, fell into the water when the deck collapsed and was never seen again.

The world of bridge engineering was shocked by this collapse, and a board of inquiry was immediately set up to look into the disaster. Two of the experts involved in the inquiry were our old friend Othmar Ammann, designer of several slender New York suspension bridges, and Theodore von Kármán, an internationally recognized expert on aerodynamics. Conclusions: bridge engineers did not fully understand the aerodynamics of wind loading; a phenomenon now known as the *von Kármán effect* was responsible for the resonant oscillations; several other suspension bridges of the Tacoma Narrows type needed to be strengthened, pronto.

Whoa, let me back up and explain what is going on here. The von Kármán effect is familiar to you in a couple of everyday manifestations. Venetian blinds flutter in the wind, and reed instruments make sounds when air is blown over the reed. For suspension bridges of the Tacoma Narrows type a diagram may help; see figure 5.20. The Tacoma Narrows bridge deck was of a very slender and elegant plate girder construction. Because of its great length and narrow width, the deck was particularly flexible, and this flexibility permitted resonant oscillations as explained in

28. You can search for "Tacoma Narrows" online and see this short film on YouTube, Wikipedia, and a number of engineering websites. Galloping Gertie was designed to withstand a steady 120-mph wind loading but collapsed in a 40-mph wind. Wind-induced resonance is not quite the same as the mechanical oscillations we have seen so far because the wind force is not itself periodic. Instead, it induces periodic motion at the bridge deck's natural frequency. As you may infer from the fact that this effect was not understood in 1940, the detailed physics of wind-bridge aerodynamics is much more complicated than that of mechanical resonance.

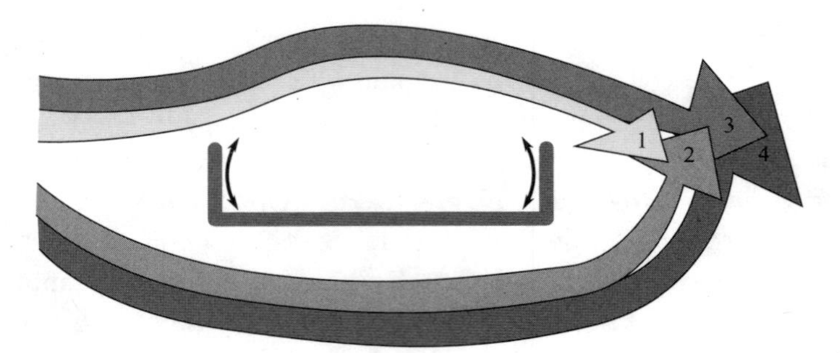

FIGURE 5.20. Wind loading can induce torsional oscillation in a bluff body such as a plate girder suspension bridge deck (seen here end-on). The wind passes over and under the deck, giving rise to eddies on the leeward side that tug at the deck, causing it to oscillate from side to side. The oscillations increase the area of deck exposed to the wind, which increases the downwind eddies, which increase the oscillations. This vicious cycle can persist until the bridge deck is shaken to pieces.

figure 5.20. A flurry of technical papers on the dynamic effects of wind loading upon suspension bridge decks was published soon after the disaster. The trend towards very slender and light suspension bridge decks ended at Tacoma Narrows. Subsequent recommendations to bridge designers included the following rules of thumb: the deck of a suspension bridge should have a width that is at least 1⁄30 of the span, and a depth in the range 1⁄50–1⁄90 of the span (Tacoma Narrows was 1⁄350). Truss decks were less vulnerable than plate girder decks because the open structure permitted wind to pass through, rather than go around, reducing the damaging wind eddies. Closed plate girder decks should be designed to be more aerodynamic, like an airplane wing, which would also serve to reduce eddies downwind.[29]

There were, at the time of the Tacoma Narrows collapse, half a dozen new suspension bridges in the United States that required stiffening as a result of the inquiry. One of these was Ammann's Bronx-Whitestone bridge. Originally, this bridge had a plate girder deck with no stiffening. Worryingly, it had displayed wind-induced oscillations of 2–3 feet amplitude. The addition of a stiffening truss along the length of the deck may have

29. The Forth Road Bridge, built a generation later than the Tacoma Narrows, would be designed with longitudinal vents in the bridge deck between the lanes of highway so that air could pass freely between them and so reduce the pressure differences that give rise to eddies and resonant oscillations. Recall that the Forth Road Bridge design was thoroughly tested in a wind tunnel prior to construction.

compromised bridge elegance and aesthetic appeal, but bridge engineers (and certainly bridge crossers) have slept more easily as a consequence. Another bridge built at the same time and in the same style as Tacoma Narrows is the Deer Island Bridge, south of Bangor, Maine. This bridge (half the length of its collapsed sibling on the other side of the continent) still stands today. The original bluff-bodied, solid girder deck suffered severely from wind-induced oscillations. Fairings were added on the sides of the deck to make it more aerodynamic, reducing the leeward eddies. Diagonal cable stays now stiffen the deck further (just as the masts of a sailing ship are stiffened against the wind by stays), at the cost of giving this bridge a cats-cradle appearance. In addition, nowadays there are 18 wind sensors and accelerometers along the deck to tell engineers about wind conditions and about the bridge's response to the wind.

The replacement for Galloping Gertie, another suspension bridge that was completed in 1950, was overdesigned to reassure the public, and it has proven to be safe.

Bridges and Earthquakes

Finally, I will say something about another type of retrofit that existing bridges are undergoing as a consequence of our increasing understanding of bridge dynamics under complicated loading. The standard procedure when building a bridge—whatever type of bridge it may be—is to anchor the piers firmly to the underlying bedrock. Sensible under most conditions, but during an earthquake this may not be the best approach, engineers now realize. If you are riding a bucking bronco, it might not be such a good idea to hang on tight because then you will feel every single jolt. If you hold on rather more gingerly, keeping some distance between you and the saddle, then you can smooth out the violent movements and emerge relatively unscathed. Similarly, current thinking is that it is probably better to permit bridges to "dance" during an earthquake than to hold them fast to the violently shaking bedrock. What this means in essence is: loosen the bolts that hold the bridge down.

Sounds counterintuitive, doesn't it? Yet this is precisely what has been done to the Lion's Gate Bridge in Vancouver, Canada. A retrofit in 2000–2001 included an upgrade to improve this suspension bridge's survivability during an earthquake. After detailed multimillion-dollar studies and countless technical reports, the bridge engineers decided that the best thing they could do to improve the chances of the bridge surviving an earthquake was

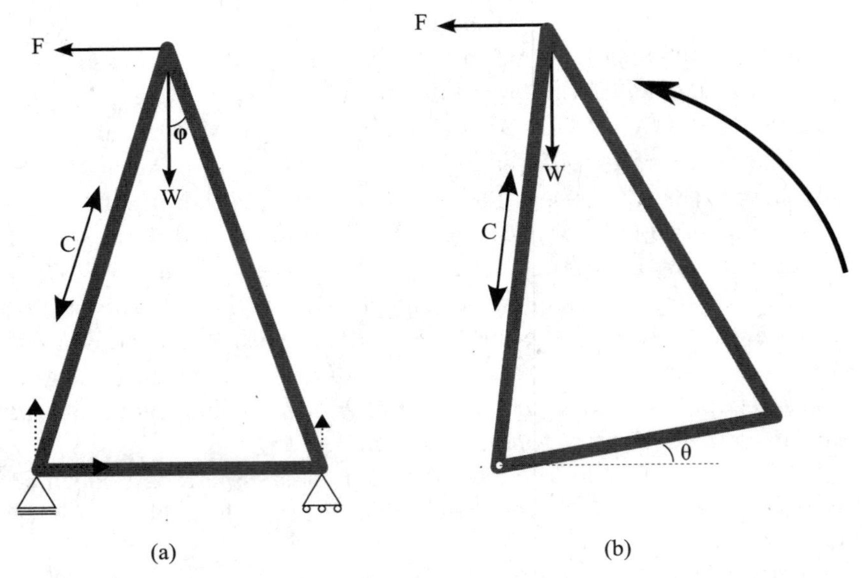

FIGURE 5.21. Triangular model of a pair of bridge piers holding up a deck of weight *W* at the apex. (a) The piers are fixed to the ground in the usual way. A sudden lateral load *F*, due to an earthquake, affects the compression stress, *C*, of the left pier as shown in appendix note 10. (b)The piers are allowed to rock on their foundations. This substantially reduces the compression stress due to the same force *F*.

to loosen the bolts. As one engineer wryly expressed it, "Fifty dollars to take the nuts off and two million dollars to prove it works." You can see his point. From an engineer's perspective, suggesting this course of action is risky. The natural reaction of most folk to such a suggestion is to think that the engineer has spent too much time being shaken up by earthquakes—the only loose nuts are in his head. Imagine if he is wrong, and the bridge collapses after the bedrock attachments have been intentionally weakened: he will be blamed for the collapse. In fact, the detailed multimillion-dollar studies recommended other changes as well, such as improved bracing by adding extra trusses to strengthen the bridge piers, but the main recommendation was this bizarre notion of loosening the pier connections to bedrock. The earthquake scenario for the upgrade was a "475-year event," so we are not talking minor tremors here.[30] Let us see if we can understand

30. An "*n*-year event" is a natural disaster (flood or earthquake) that is expected to occur once every *n* years. For such disasters, larger *n* means increased severity.

how this strange course of action (which has now been implemented—probably the first repair of its kind in the world) might conceivably work.

No prize will be awarded to the reader who has guessed that my approach to gaining some insight into this problem will be via a simple truss model. If I model a pair of piers that hold up the Lion's Gate Bridge by a triangle, as in figure 5.21a, then I can determine the stress in each pier by truss analysis. You can see from the graphs in figure 5.22 that the compressive stress is much less for the rocking bridge than for the rigidly fixed bridge. The calculations are outlined in appendix note 10.

Of course, the physics of the real bridge is far more complicated than this model suggests, but you can see how the recommended seismic upgrade idea is not totally, well, nuts. Care must be taken to ensure that a

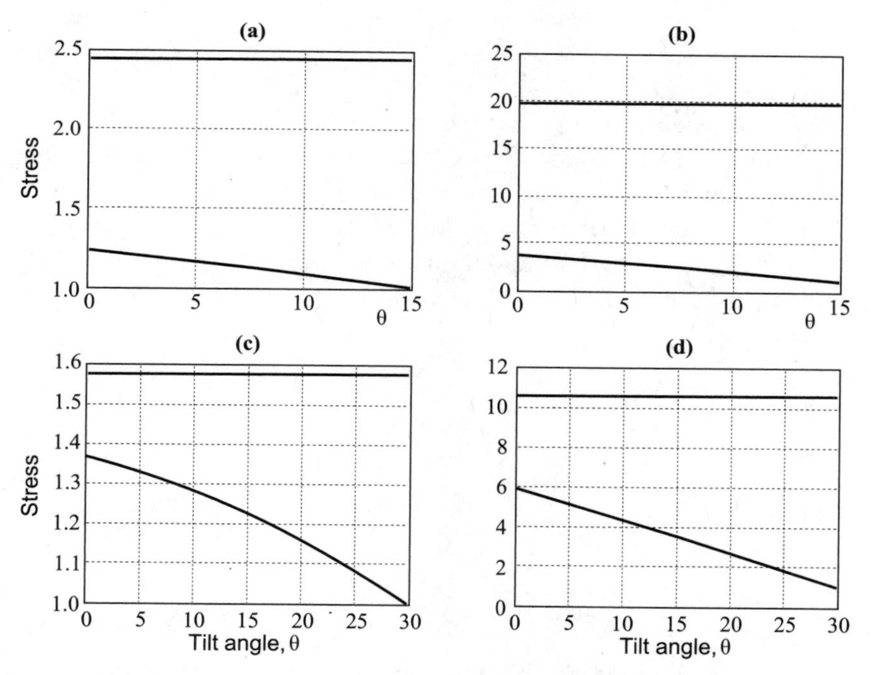

FIGURE 5.22. Compression stress in the left pier of the bridge model of fig. 5.21b versus tilt angle θ. For comparison, the stresses that arise in a fixed-bridge model (fig. 5.21a) are shown as horizontal lines. In the four graphs the bridge pier angle φ and the ratio $R = F / W$ of earthquake lateral force to bridge weight is as follows: (a) $(\varphi, R) = (15°, 1)$. (b) $(\varphi, R) = (15°, 10)$. (c) $(\varphi, R) = (30°, 1)$. (d) $(\varphi, R) = (30°, 10)$. In all cases the rocking bridge experiences less stress, particularly for strong lateral forces.

rocking bridge returns to its normal position gently because a hard landing might damage the bridge. Viscous dampers are installed at the base of the piers to absorb energy. One reason that real bridge physics is more complicated is because the Lion's Gate Bridge has 24 piers of different lengths (because of the height of the bedrock), and these piers are of course connected by the stiff bridge deck.

SUMMARY: Large-span bridges are of different construction from smaller bridges: beams and arches give way to cantilever, suspension, and cable-stayed designs. (Bridges with specialized load or design requirements—for example, those carrying railways and requiring a curved deck—may require special designs, such as the economical trestles or box girders.) Cantilever bridges exert no horizontal forces and thus require no abutments; however, they are massive. Suspension bridges require strong abutments to resist the strong horizontal forces that arise from this design; however, the forces are easy to analyze and decks can be light. The flexibility of early suspension bridge decks gave rise to damaging resonance oscillations due to dynamic wind loading. These problems were solved by stiffening the decks and making them more aerodynamically stable. Cable-stayed bridges are intermediate in structure between cantilever and suspension bridge designs and have some of the advantages and disadvantages of both. They are difficult to analyze and require modern technology to implement successfully.

CHAPTER 6

Dam It

Dam Classification

With the exception of the Great Wall of China, dams are the largest of human construction projects. I will say some more about the stupendous scale of these structures in the next section, but before I can discuss dams with you sensibly, I need to bring you up to speed on the different types of dams that can be found today. It should not come as too much of a surprise to learn that there are several ways to build dams, depending upon the nature of the rock foundation and valley walls, the required dam length and purpose, and economics, yet few people stop to think about it. We tend to be awed by the scale of these structures, and only when visiting a large dam might we wonder about the shape. In this section I will go through the four basic types of dams, to show you what they are and why. Simplified diagrams of these types are provided in figure 6.1.

Embankment Dams

Embankment dams are the most common type of dam in most countries of the world. An embankment dam is illustrated in figure 6.1a. The simplest embankment dams consist of a lot of dirt piled up across a river bed or along a shoreline to restrict water flow. Because of the shape of these dams, water pressure tends to push them into the ground, so that there is little danger that they will tip over. Embankment dams are the only type of dam not made of concrete today; they are formed from rock fill or earth fill. They have two advantages over other dam types. First, the construction material is obtained locally and is straightforward to put in place, which keeps down the cost of construction. Second, the large base area of the dam means that there is relatively little pressure upon the foundation (compared with the other types of dam), and so embankment dams can be built

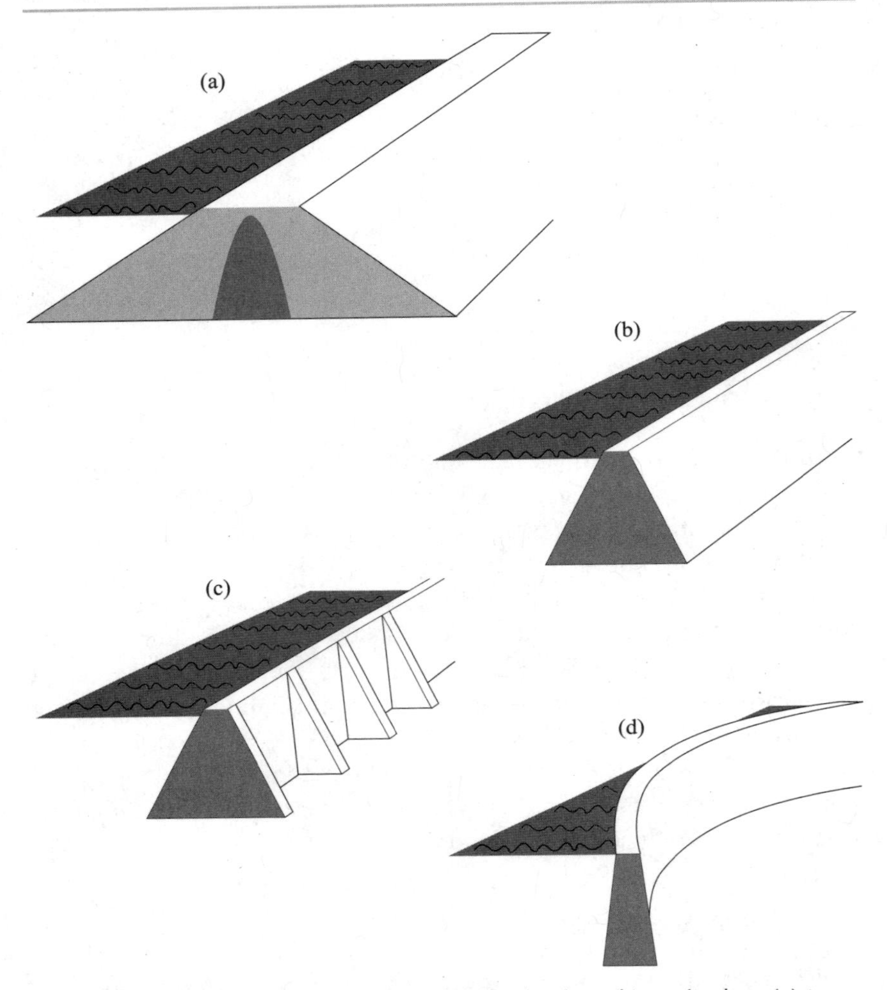

FIGURE 6.1. Four basic types of dams: (a) embankment dam; (b) gravity dam; (c) buttress dam; (d) arch dam. Embankment dams are usually constructed from earth or rock fill; the others are concrete.

where other types cannot. They are good for wide valleys, over open-jointed (loose, porous) rock—because of their relatively low weight per unit area—and are not particularly susceptible to earthquake damage.

The main issue with embankment dams is that, because of their construction material, they are water-permeable. This problem can cause erosion and eventual destruction of the dam unless addressed. Indeed, in the

United States before 1930 these embankment dams had a poor record. They tended to collapse, usually as a result of overtopping during flood conditions; once water started to flow over the dam, it would carry away the rock fill or earthen core. Modern embankment dams are constructed with vibrating rollers to compact the soil and gravel; more importantly, they are provided with an impermeable face or core. That is to say, an impermeable material such as dense clay is placed on the upstream face of the dam to prevent seepage through the dam structure or a core of impermeable material is placed within the dam. Much thought has been devoted to the drainage from embankment dams, with the seepage of different grades of construction material carefully measured.[1] Materials that are resistant to seepage ensure the stability of the structure, preventing erosion, liquefaction, slipping or sliding when water overtops the dam.

The advantages of embankment dams usually outweigh the disadvantages and so, as we will see, many of the largest and most famous of the world's dams (such as Aswan High Dam in Egypt) are of the embankment type.

Gravity Dams

Gravity dams rely upon their weight to resist water pressure. Today, they are constructed of concrete (though historically they were masonry-built) and can be hollow or solid. As you can see from figure 6.1b, gravity dams tend to be less massive than embankment dams, but they can still be huge. Water pressure, rather than acting to push the dam into the ground, instead tries to push it over; the weight of the dam resists this overturning torque. One advantage of gravity dams is that they are of quite simple design, usually straight and with a constant cross section along the length. The main disadvantage is cost. Vast amounts of concrete are required for gravity dams; indeed, construction of the largest such dams requires a cement plant to be built on site. For this reason, gravity dams tend to be shorter than embankment dams, extending across narrow gorges. Another factor that limits the popularity of gravity dams is the stringent requirement they place upon the bedrock foundation. The bedrock must be strong enough to

1. Thus, for example, gravel permits 10,000 times more water seepage than sand, and 10 million times more than silt. Dense clay is up to 10 times better than silt at resisting water seepage.

support the dead weight of all that concrete. It must also be impervious, so a careful geological assessment of the foundation bedrock must precede any gravity dam construction.

Because gravity dams look sturdy and reassuring, the general public likes them. A solid dam is important for those people who live immediately downstream of a large dam in flood-prone regions.

Buttress Dams

Buttress dams resemble gravity dams but are much less bulky. The saving on concrete is a distinct advantage, but because they are less strong, they must be reinforced by adding buttresses to the downstream side of the dam (fig. 6.1c). These dams are harder to design and build than the embankment or gravity types, and they cannot be built so high. On the other hand they are more versatile—they are a compromise design in many ways. They impose less stringent restrictions upon foundation bedrock than do gravity dams (from which they evolved), because of their reduced weight, and they require less material than either embankment or gravity dams.

Arch Dams

Of all the dam types, *arch dams* have the highest *wow!* rating, in part because they are built in dramatic landscapes of deep gorges. Arch dams are the thoroughbreds of the dam world—high performance but very demanding. As you see illustrated in figure 6.1d, arch dams are curved, with the outside of the curve upstream. The dam ends are braced against strong rock walls. So, an arch dam acts like a structural arch on its side, resisting horizontal water pressure in the same way that vertical arches resist the weight of a building or bridge. Arch dams make use of the strength of concrete, rather than its weight, to stay in place. In fact, as with arches in buildings, the force that they are designed to resist helps to compress the arch. This is an advantage for dams because they need to be impervious to water leakage. Again, as with building arches, arch dams transfer large forces to the abutments, which in this case are the rock walls. So immediately we see the main *disadvantage* of arch dams: they can be built only in locations where there are steep valleys, canyons, or gorges with strong rock walls (and a strong bedrock foundation).

True arch dams can have relatively thin walls because of the strength of arches, yet one of our most famous arch dams is in fact a hybrid arch-

FIGURE 6.2. Hoover Dam, a gravity-arch design. Note the strong canyon walls against which the arch is braced. The four towers in Lake Mead are the intakes for the hydroelectric plant, which is located at the foot of the dam. Image from Wikipedia.

gravity dam. The arch of Hoover Dam is plain to see (fig. 6.2); the gravity part is evident in the tapered cross section—thick base and narrow crest at the top.

Living with Dams

We build dams to serve local (and distant) communities in several ways. Most obviously, dams are built to hold water[2]—to prevent flooding and to furnish a reservoir of water for domestic and industrial consumption and for irrigation. Given that a dam backs up a large head of water, it would be a crying shame if all that potential energy was not put to good use, which brings us straight to the second main reason for dam construction: hydroelectric power. In many countries of the world, this is the overriding *raison*

2. Technically, dams are built to retain water rather than to prevent the flow of water. If flow prevention is the main purpose, then the construction is known as a dike or levee.

d'être for large dams; the electrical power output of the giant dams is awesome, as we will soon see.

A less well-known advantage of dams—rarely a sufficient reason on its own for dam construction but often a not inconsiderable benefit—is that the backed-up water increases the navigable stretch of a waterway. The reservoirs that a large dam creates can stretch for hundreds of miles (Lake Mead, created by Hoover Dam, is a well-known example). Because the dam can let excess water through spillways, the upstream water level can be controlled, and so ships and boats are able to navigate such waterways year round with confidence, without having to worry about high and low tides, seasonal flood surges, and so on.

A large dam project has a significant impact upon people upstream and downstream for hundreds of miles, not only after it is completed but even before the work has begun. Consider just one aspect of dam impact, that of water flow management. Lake Mead, above Hoover Dam, covers nearly 250 square miles. Grand Coulee Dam in Washington State pumps water up to arable land covering several hundred square miles. Aswan High Dam in Egypt backs up the Nile River to create Lake Nasser, one of the largest reservoirs in the world (300 miles long and 10 miles wide). Three Gorges Dam in China, recently completed though not yet fully operational, creates a reservoir that is 400 miles long. It will control flooding on the volatile Yangtze River, which, historically, has been the cause of many hundreds of deaths,[3] and it will permit an extra 50 million tons of freight each year to travel the Yangtze, aided by large navigation locks that will be able to raise ships 370 feet.

Yet all these benefits come at a price that causes concern in some quarters. The land that is now under Lake Mead may have been sparsely populated, but the same cannot be said of the banks of the Yangtze: more than one million people have been relocated to make way for Three Gorges Dam reservoir. The Nile floods are now controlled and Egyptian agricultural output has doubled as a consequence of Aswan High Dam, but prior to construction thousands of people were displaced, during construction a massive and expensive international project moved important archaeological monuments (including the temples at Abu Simbel) to higher ground, and since completion the lack of floods has caused great upheaval among

3. Fifteen hundred people were killed in a "5-year" flood in 1998. Three Gorges Dam will be able to control much larger inundations from "1,000-year" floods. Flood management is a significant reason for the construction of this dam.

the traditional river delta farmers, who for thousands of years had relied upon the fertilizing silt water that inundated their land each year. Numerous environmental groups and tourist organizations around the world object to many dam projects; they urge caution and demand careful environmental impact studies ahead of construction. The main objections are that a new, large dam displaces whole communities and can result in the loss of arable land. It may also detract from the natural beauty of a region, inundate fish spawning grounds, inhibit the seasonal migrations of fish, and threaten species. Dam reservoirs may inundate archaeological sites and burial grounds. Some claim that these reservoirs can create unstable hillsides and that the large "artificial" weight of a dam can lead to earthquakes.

Proponents of these large dam projects are increasingly aware of the potential environmental impact (indeed, some would say that the main purpose of many dams is to make an environmental impact) and argue the case for proceeding with the project. Three Gorges Dam has come in for particular criticism, mostly because of the large number of people who have been obliged to relocate. The Chinese government claims that the loss of traditional agricultural land to the reservoir is small because of the steep sides of the gorges. Clearly, there are winners and losers with a construction project that alters hundreds of square miles of topography, especially in traditionally populous areas. A powerful incentive for many large dam projects across the world is that they increase the available national electricity supply significantly. When fully completed, Three Gorges Dam will add more than its present 18,200 megawatts to the Chinese national grid; this is enough electricity to power a medium-sized country.

The construction and operation of large dams requires auxiliary works that impact the local economy and environment for better or worse. Quarries or cement plants, and sometimes even dedicated steel plants, are built on site. Accommodations for thousands of workers, as well as roads for construction equipment, have to be provided. Provision must be made for evacuating silt and flotsam (floating logs and garbage, for example) that build up in the dam reservoir. In many cases locks are built to facilitate shipping movement. Intake structures are built near the dam to take water into a power station. Spillways, gates, or valves are needed to control the discharge of surplus water, while canals or pipelines may be required to distribute water over hundreds of miles. A simplified plan for a dam plus auxiliary works, including hydroelectric turbines, is shown in figure 6.3.

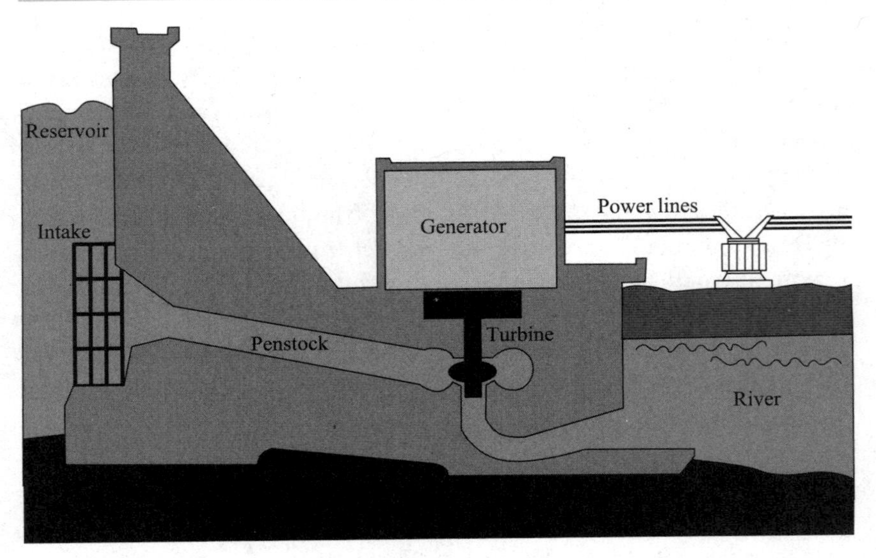

FIGURE 6.3. Basic elements of a hydroelectric dam. Adapted from an image by the Tennessee Valley Authority.

Historically, dams have impacted communities by failing catastrophically. This is not necessarily because of some construction flaw or because of a misunderstanding of the physical forces that need to be taken into consideration. (I will discuss these forces in the last section of this chapter.) Dam engineers in the nineteenth- and twentieth-century United States knew as much about building dams as anyone on the planet, but even here there were many significant failures. In 1889 South Fork Dam near Johnstown, Pennsylvania, failed, causing the deaths of 2,000 people downstream. In 1928 the failure of St. Francis Dam near Los Angeles killed hundreds of people. In 1976 Teton Dam in Idaho failed, killing 14 people and causing a billion dollars of damage. There have been other dam failures since: the average failure rate (mainly in small dams) is about one per 10,000 dam-years. Given that there are about 77,800 dams in the United States, this figure suggests that there are seven or eight failures per year. Most of these are due to exceptional flood water levels and occur with small dams where the failure causes little damage, but they naturally leave the public a little uneasy. We expect our engineering projects to work; failure indicates that construction engineers have not taken everything into account, and disaster movies remind us of the damage that a dam failure can do.[4] To ease public concern—about environmental impact as well as safety—many large dams encourage the public to visit and to be educated about the project. Such is our interest in these behemoths of construction that the dams become tourist attractions in themselves. Hoover Dam, for example, draws 7 million visitors each year.

We will see in the next section that Hoover Dam is not the world's biggest, though it was when first completed. (Today, it is not even the biggest dam in the United States.) The story of its construction is exceptional and yet typical of these enormous engineering projects, and so I will spend a few paragraphs describing the building of Hoover Dam. This history is well documented and is instructive. Begun in 1931, construction was always much more than just another Depression-era work creation project. When completed, the dam—originally called Boulder Dam—would be 60% higher and 2½ times larger than any dam in existence; it would be the world's largest concrete structure and would contain the world's largest electric power facility.

Seven thousand workers, some with their families, flocked to the site on the Arizona-Nevada border to obtain work. Thousands more, less fortunate, were turned away. Working conditions were initially very bad, both on and off site. The heat in summer and the cold in winter had to be endured in tents initially, and then in the notorious Ragtown, before improved accommodations were provided in what would become Boulder City, Nevada. The work was often dangerous: over the years of construction 112 workers lost their lives. Despite the conditions this project was completed in 1936, two years ahead of schedule, spurred by President Hoover, who took a personal interest in the dam construction.

Construction proceeded as follows. Two temporary dams were built, upstream and downstream of the dam site, to form a *cofferdam*. Their construction required one million cubic yards of earth fill and rock fill to be put in place. Water from the Colorado River was diverted through four tunnels blasted through the walls of Black Canyon—two tunnels on the

4. One such movie was *The Dam Busters* (1955), a fictionalized account of a true World War II air raid by the 617 Squadron of the British Royal Air Force on the Möhne, Eder, and Sorpe dams in Germany, the destruction of which was intended to cause the industrial Ruhr region to flood and so wreck a significant sector of German industry. Specially designed bouncing bombs eventually breached two of these dams, causing widespread flooding, but the benefit to the Allied war effort of this raid is debatable.

FIGURE 6.4. A 1941 photograph of Hoover Dam, from downstream. Note the tunnel outlets. Image courtesy of the National Archives and Records Administration (ARC 519837).

Nevada side and two on the Arizona side, with a combined length of 3 miles. To create these tunnels, 1.5 million cubic yards of rock and dirt was removed. Each tunnel was 50 feet in diameter and lined with 3 feet of concrete. When the tunnels were completed and water flowed through them (see fig. 6.4), work could begin on the main dam foundations. Another 2–3 million cubic yards of material was removed to reach bedrock, which lay 135 feet below the original river bed. Once bedrock had been reached, the concrete could be poured.

Pouring concrete began in June 1933 and took 2 years. Over 5 million barrels of cement were required to make the 3.25 million cubic yards of concrete that was needed.[5] When concrete cures, it generates heat. For a

5. This is about the same volume as the Great Pyramid of Giza. It is claimed to be the amount of concrete that would be required to build a two-lane highway from San Francisco to New York. Or, if you prefer, it is the volume of concrete that you would need to build a sidewalk from the Arctic Ocean through North and Central America

normal-sized project, such as a house foundation, the heat generated is negligible, but for Hoover Dam it was enormous—so much so that 582 miles of steel pipe was required, with ice water flowing through it, to dissipate the heat. Without this active cooling mechanism, the concrete would have taken over a century to set properly, and the heat stress would have caused fractures. The concrete was poured, not as a single monolithic block, but as many tall, interlocking columns that resembled a giant jigsaw puzzle from above. These columns would be pushed together and held in place by the pressure of water at the upstream face of the dam. Auxiliary structures were built,[6] and the first of 17 turbines was installed and became operational in 1936. The final turbine would not be operating until 1961, raising the total power output to 2,000 megawatts.

Despite its epic scale Hoover Dam would, within a few years, be dwarfed by the Grand Coulee Dam, which required almost three times the volume of concrete. Yet, even the Grand Coulee is not the biggest dam in the world, as we will now see.

Dam Evolution and Growth

Perhaps surprisingly, all four of the basic dam types have been known to mankind for thousands of years. The first dams are thought to have been built in Mesopotamia around 5000 BC. Remains of an embankment dam dating from 2800 BC have been found in Egypt. This dam was probably 45 feet high and 370 feet long; it failed during a flood. Remains of the oldest known gravity dam, in Jordan, are about the same age. The oldest surviving dam (second century AD), constructed to divert river water for irrigation in India, is over 1,000 feet long and is made of unhewn stone; it is in excellent condition. Other embankment dams from the first millennium have been found in Yemen, and two that were built by the Romans in Spain are still in use. Arch dams were built occasionally by the Romans (they preferred gravity or embankment dams), and there is evidence that the builders understood how water pressure pushed together the dam masonry, making the structure stronger and less permeable. Arch dams were also built in Persia (modern-day Iran) by the invading hordes of Mongols that swept through much of the Old World during the Middle Ages.

---

to the southern tip of South America. Hoover dam is 726 feet high and 1,244 feet long; it is 660 feet thick at the base, tapering to 45 feet at the crest.

6. Did I mention the 72,000 tons of steel that were needed to build this dam?

Archaeology shows that in many parts of the Old World, dams were built to the same basic plans as they are today, and for much the same reasons—except for hydroelectric power, of course. Dams retained water for domestic consumption, irrigation, or flood control. Dams (or rather, dikes and levees) were constructed to divert water through waterwheels, which were the ancient forerunners of modern turbines, so even this application of dams—using water energy to create useful power—is far from being new. What is new, say within the past 150 years, is a scientific understanding of how to build dams, based upon an appreciation of the forces that act upon a dam and upon the strength of the materials that constitute a dam.

In 1850 a French engineer, de Sazilly, was the first to show that the best shape for the cross section of a gravity dam is triangular. This is the case because of the distribution of water pressure, as we will see, and had been understood empirically the previous century, judging from the shape of a late-eighteenth-century gravity dam in Mexico (by an unknown designer). In 1804 an impressive multiple-arch dam was built in India by the British. The Mir Alam Dam reservoir still supplies water to Hyderabad today. It is composed of 21 semicircular arches of various spans, with each arch curve having a constant radius. This curvature is important because we know that cylindrical or spherical shell structures are very strong (think of all those architectural vaults and domes). Other examples of these *constant radius* arch dams of the nineteenth century include Jones Falls Dam in Canada and Zola Dam in France; the latter was the first dam to be built following a detailed engineering stress analysis.

A significant step in dam evolution took place in the early twentieth century, when concrete replaced jointed masonry. Concrete proved to be a more versatile material than masonry, enabling designers to construct optimum (and more elegant) curved shapes for arch dams. Much of the development in dam technology at this time took place in the United States. Modern analysis techniques were required prior to the development of *constant angle* (a.k.a. *variable radius*) arch dams: the first to be built, in 1914, was Salmon Creek Dam in Alaska. I will have more to say about the two types of arch dams—constant radius and constant angle—in the next section. For now we merely need to appreciate that they are relatively recent innovations that reflect two developments in civil engineering: an increased theoretical understanding of forces and the widespread availability of concrete.

Though some of the oldest dams are quite large, the really big dams-on-

TABLE 6.1. Highest dams in the world and in the United States

Order	Name	River	Country or U.S. state	Type	Height (ft)	Year completed
World						
1	Nurek	Nurek	Tajikistan	Embankment	990	—
2	Grande Dixence	Dixence	Switzerland	Gravity	940	1961
3	Inguri	Inguri	Georgia	Arch	900	1980
United States						
1	Oroville	Feather	California	Embankment	780	1968
2	Cyprus	Bruno Creek	Idaho	Embankment	750	1982
3	Hoover	Colorado	Nevada / Arizona	Arch-gravity	740	1936

steroids are a twentieth-century reflection of engineering and societal advances. You can appreciate from my summary of the construction of Hoover Dam that physics and materials science are not the only skills that a nation requires before embarking upon such enormous projects; the enterprise requires huge amounts of money, organization and logistics, and national infrastructure. There is not much point in building a big dam in the middle of nowhere (and dams usually have to be built where topography dictates, often in relatively inaccessible places) if there are no roads to get there or no national electricity grid to benefit from the hydroelectric-generating capability. In much of the world, the requisite infrastructure lagged behind theoretical know-how, so that very large dams became feasible across much of the world only within the last century. Tables 6.1 and 6.2 summarize the vital statistics of humankind's largest dams.

These tables do not tell the whole story. Nurek Dam in Tajikistan was begun decades ago, but construction has been delayed due to political turmoil and technical difficulties. When (if) it is completed as planned—the current structure is only part of the planned dam—it will be the highest dam in the world and so will head table 6.1. Table 6.2 contains no Canadian entry although Canada is one of the biggest producers of hydroelectric power. (Geology plays a significant role in dam location, as we have seen: this is why Canada and Washington State are able to generate a lot of hydropower.)

TABLE 6.2. Hydroelectric dams with largest power output in the world and in the United States

Order	Name	River	Country or U.S. state	Type	Power (MW)	Year completed
World						
1	Three Gorges	Yangtze	China	Gravity	18,200	2009
2	Itaipu	Purona	Brazil/ Paraguay	Buttress/ hollow gravity	12,600	1982
3	Guri	Caroni	Venezuela	Embankment	10,300	1986
United States						
1	Grand Coulee	Columbia	Washington	Gravity	6,180	1941
2	Chief Joseph	Columbia	Washington	Buttress	2,620	1979
3	Robert Moses	Niagara	New York	Buttress/ gravity	2,500	1961
9	Hoover	Colorado	Nevada/ Arizona	Arch-gravity	2,080	1936

Dam Physics

This section addresses several aspects of the physics of dams. We investigate the shape of arch dams, and then we examine the forces that act upon a dam via a simple truss model and some other elementary Newtonian physics. For readers who are interested in a more technical aspect, I estimate in note 11 of the appendix the water flow required through a hydroelectric dam to achieve the stated power output of the dam.

Arch Dam Shape

In the preceding section, on dam evolution, I mentioned that two different shapes of arch dams have evolved: the constant radius arch dam and the constant angle arch dam. The earlier of the two designs was the constant radius dam (fig. 6.5a). The basic idea was to make a dam wall have the shape of part of a cylinder, cut out to fit the valley.[7] Even thin-shelled cylinders can be strong, as we saw with vaults, and so we can see why an

7. "The valley of the dammed," so to speak.

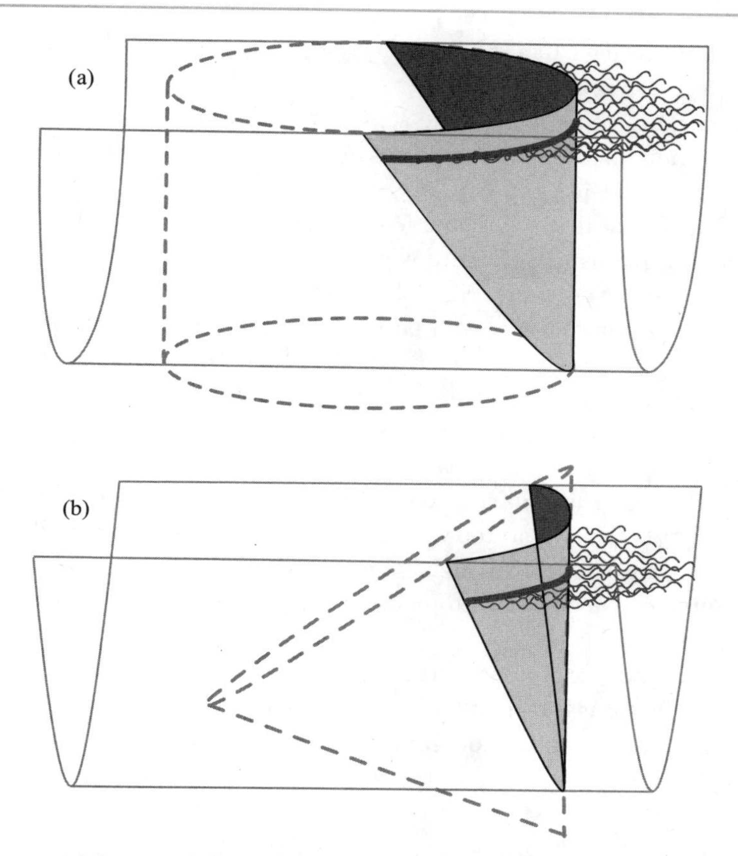

FIGURE 6.5. (a) Constant radius arch dam. The dam wall is a section of a cylinder (dashed lines). The wall foundations are in bedrock on the sides and the bottom of a U-shaped valley (gray lines) where the valley intersects the cylinder. The dam face is shown as shaded areas. (b) Constant angle arch dam. The dam wall is a section of a cone (dashed lines). The wall foundations are in bedrock on the sides and the bottom of a V-shaped valley (gray lines) where the valley intersects the cone. Again, the dam face is shown as shaded areas.

arch dam that is of constant radius at all heights will be able to resist the pressure exerted by water upstream (assuming, as always for arch dams, that the valley walls are strong). Arch dams are strong because concrete is strong in compression; the amount of concrete needed for a constant radius arch dam is much less than for a gravity dam covering the same valley.

If the valley is V-shaped, rather than U-shaped, then a constant angle arch dam is the better choice. This structure is illustrated in figure 6.5b. Here it is the subtended angle that is kept constant at all heights, while the dam radius is allowed to vary. You can see from figure 6.5b that the resultant dam shape is quite complicated (and so concrete is a much more convenient construction material than stone). Why bother? There are two reasons.

First, think what the bottom section of a constant radius arch dam would look like if it spanned a V-shaped valley rather than a U-shaped valley. The radius of curvature would be large compared to the narrow span, so that the curvature would be barely noticeable. The strength of the curved cylindrical shell would be lessened; there would not be enough space for the curve to develop—for the arch to brace itself against the pressure of water. It is much better to have a smaller radius at the bottom of such narrow valley floors, where the water pressure is greatest, so that the benefits of curvature can be realized. This is what the constant angle arch dam achieves. The radius of curvature for the dam gets smaller lower down, fitting the dam to the valley wall in such a way that the subtended angle is always the same, at all heights. The shape is quite complicated, and difficult to convey on a two-dimensional page, but I hope that you get the idea from figure 6.5b.

The second reason that constant angle arch dams are better than constant radius dams—at least for V-shaped valleys—is that they require less concrete. The volume of concrete required is minimized if the angle subtended by the arch is 133°; for this angle, the concrete needed is only 43% of the volume required for a constant radius dam covering the same valley.[8] In practice, it is not always possible to attain the optimum angle for all dams of this type, but they generally lie in the range of 100° to 140°.

Forces That Act on a Dam

In this section we will consider the forces that act upon a dam and how engineers take them into account when designing a dam. As in previous chapters, a truss model can give us an idea, at the expense of only a little

8. You may think that the cost saving is not a big deal, given that concrete is a relatively inexpensive material. However, the volumes are so large for these big dams that the difference in construction materials can represent an enormous saving. Where arch dams are not feasible and gravity dams are required, it may still be possible to cut down on the volume of concrete needed (recall that gravity dams require prodigious volumes). The Allt na Lairige Dam in Scotland makes use of post-tensioned steel rods on the upstream face of the dam—giant rebar, if you like—to reduce the volume of concrete required by 40%.

calculation, of the forces that act within a structure. Then we will look specifically at gravity dams to see what forces dam engineers have to take into account and how these forces influence the shape of the dam.

A BRIGADE OF FORCES

The forces that need to be considered arise from several sources. Here is a summary:

1. *Dead load.* Concrete weighs about 150 pounds per cubic foot, and so the weight of the dam is considerable; for gravity dams, it is the weight that holds back the water. The bearing strength of foundation bedrock must be enough to support this load.
2. *Headwater pressure.* The hydrostatic pressure of the water, multiplied by the wetted area of the upstream face, determines the water force against the upstream face of the dam. Dam construction must be sufficiently robust to resist this force. For some dam designs, the force on the downstream face arising from tailwater pressure also needs to be considered.
3. *Uplift force.* For a gravity dam, the headwater force will act to tip the dam over backwards, thus lifting the foundation on the upstream side. Water can creep underneath the foundation, further weakening the connection between dam and bedrock. Uplift force can be a significant problem for porous foundations.
4. *Temperature.* We have seen that the concrete curing process is an exothermic chemical reaction—it generates heat. Also, solar heating on the downstream face can set up temperature gradients. Both cases give rise to internal stresses within the concrete that can lead to cracks unless measures are taken to prevent this.
5. *Earthquake loading.* The dam must be strong enough to resist earthquakes of a specified magnitude. Engineers must ascertain how the underlying bedrock is going to react to an earthquake: will it slip, liquefy, or remain unchanged?
6. *Wave pressure.* Water waves will pound against intake towers or spillway gates, which must be made strong enough to resist these forces, year in and year out.
7. *Erosion forces.* Spillway shape must be carefully chosen to minimize the erosive effects of millions of tons of water pouring over the spillways. Similarly, the bedrock immediately upstream and downstream of the dam may need protective aprons to counter erosion.

TRUSS DAM

Figure 6.6a depicts a triangular truss model of a gravity dam in cross section, with the vertical face upstream. Our old trick, of estimating the stresses that act upon the structure by modeling it as a truss, will prove fruitful. As always, of course, we must be cautious when interpreting the results of this approach because of the simplifying approximations that it entails. Note that the weight, *W*, of the dam is distributed unevenly on the two lower pins, which represent the dam foundation sitting on bedrock. Because of the asymmetrical weight distribution (due to cross-section geometry), ¾ of the dam weight is at the upstream face and only ¼ at the downstream side. The force, *F*, that is exerted on the dam by the water is also asymmetrical; 1⁄7 acts at the top and 6⁄7 at the bottom. Why this particular division? Water force increases as the square of water depth; from this fact it is possible to show that the upper half of the dam feels a force 1⁄7 *F* whereas the lower half feels 6⁄7 *F*. Truss models permit forces only where there are pins, you may recall, and so we have to model the continuous distribution of force over the upstream face of the dam by two forces, here acting at the top and the bottom.

Of course, this approach is a significant approximation, but it will get us where we are going. We do the truss calculation and determine the three forces $T_{1,2,3}$ that act upon the three truss members. I could write down the results, but they are not so important and so will skip them—work it out yourself if you need to know the details. Suffice it to say that the horizontal member representing the foundation of the dam experiences a small tensile force.[9] The sloping downstream face is under a much larger compression force, but this is not a problem because concrete is strong in compression. This compression force on the downstream side is large for steep slopes and small for gradual slopes, which explains why embankment dams have gradual slopes: embankment dams are earth-fill or rock-fill structures and not as strong as concrete. The vertical upstream face of this dam is under a strong tensile force. This is a worry because concrete is not so strong under tension and explains why some concrete dams have prestressed steel bars inside the upstream face to strengthen the dam where it needs it (see footnote 8). This strengthening means that less concrete can be used in the dam construction—in other words, the angle φ, shown in

9. To be more accurate, a small force compared with the other forces I will discuss, but a huge force on the human scale. You wouldn't want to have to support it.

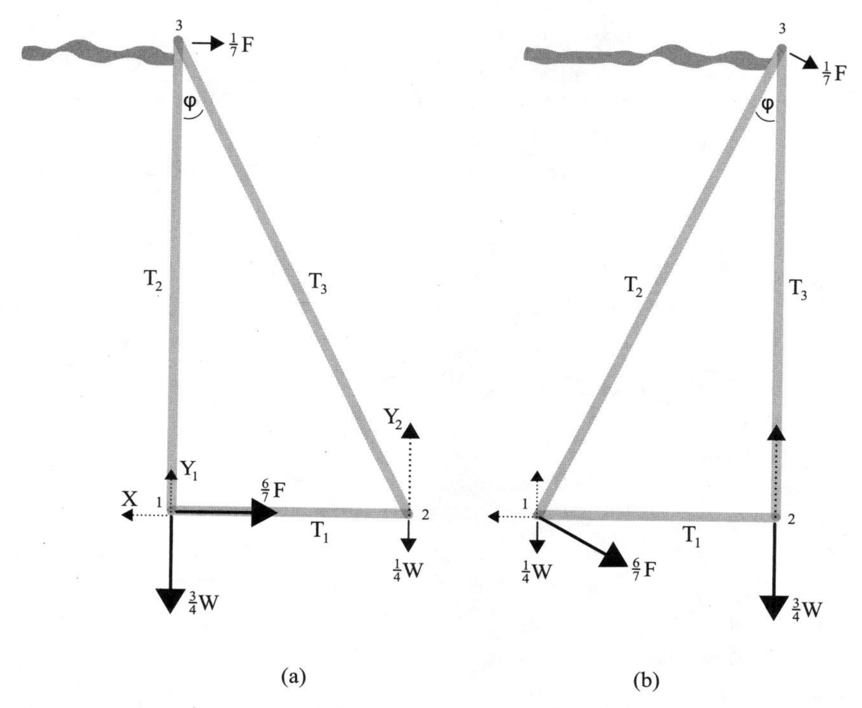

FIGURE 6.6. Triangular truss model of a gravity dam. (a) Model with sloping face downstream. The external forces that act upon the truss are water force, F, and dam weight, W, distributed among the truss pins as shown. The ground reaction forces are shown as dotted arrows. (b) The dam turned around: now the sloping face is upstream. The external forces are distributed differently, but the forces $T_{1,2,3}$ that act within the truss are much the same.

figure 6.6, can be reduced. When the angle φ is large, we find that the tensile force is not so severe, and so strengthening bars may not be needed, at the cost of more concrete.

All these results accord with common sense and with experience. Our model is giving us sensible results. In note 12 of the appendix, I push the model a little harder and find that there is a minimum allowed angle φ of 16°; smaller angles lead to uplift of the dam base, which would be bad news. Once again we have obtained results from a simple truss model that have shed some light upon what is a complicated problem. More accurate and detailed calculation of forces that act upon this shape of gravity dam are much more complicated. If we turn our triangular dam around so that the sloping face is upstream, as in figure 6.6b, the forces change as shown, but interestingly, the results do not change much. The forces that act upon the upstream and downstream faces of the dam are the same as before. The force T_1 at the foundation is now zero, and the minimum value of φ to ensure that there is no uplift force at the upstream edge of the foundation is now 17°.

TRAPEZOIDAL DAM

Real gravity dams are more often shaped like that of figure 6.6a because this triangular orientation is harder to tip over, but other shapes are possible. We will now look at a trapezoidal cross section. We will find, as in the case of our triangular model, that the physical forces acting upon the structure constrain its shape. Consider the gravity dam cross section shown in figure 6.7. On the left, water pressure is represented by arrows. The pressure increases linearly with depth. Does this pressure profile act to tip the dam over backwards, to tip it forwards, or to push the dam backwards without tipping? Imagine a brick standing on end. If you push it near the top, it will fall flat—tip over backwards; if you push it hard near the bottom, it will fall forward; push it somewhere in between and it will not tip but will slide backwards. We can show that for the gravity dam case acted upon by water pressure, the water force will try to tip the dam backwards if the center of gravity is higher than ⅓ of the water height ($x > \frac{1}{3}H$ in the notation of fig. 6.7) and will try to tip it forwards if the center of gravity is lower than ⅓ of the water height. We don't want either of these to occur,[10] and so we design the dam cross section in such a way that the center of mass is situated at ⅓ of the height. For such a design, the water force tries to push the dam but not tilt it.

Does this requirement for center of mass location place any constraints upon the shape of our trapezoid? Yes, it does. It is not difficult to show that, if the center of mass is ⅓ of the height, then the crest of the dam must be ⅐ of the base[11] ($d = \frac{1}{7}b$ in the notation of fig. 6.7).

In figure 6.7 we see three vectors: F_W represents water force, W represents the downward force due to gravity, and F is the sum of these two

10. Technically, we don't want any *torque* acting upon the dam. In particular, we don't want any torque that will act to tip the dam backwards because this will lift the front of the foundations, and we have seen that such uplift forces can be bad news.

11. We obtain this constraint by observing that, if the center of mass is at one-third of the dam height, then half the area of the trapezoid must be in the lower third of the dam, and half in the upper two-thirds. The rest is elementary geometry.

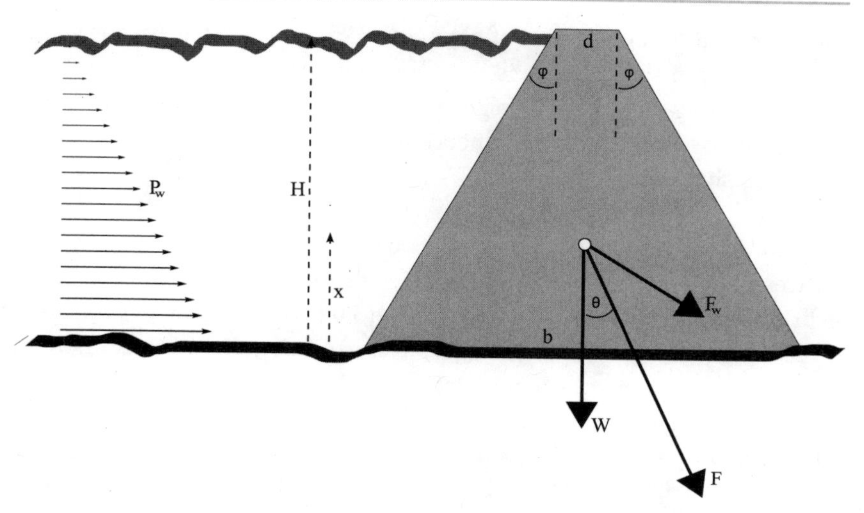

FIGURE 6.7. The pressure gradient, shown on the left, creates a force, F_W, on the trapezoidal dam that acts perpendicular to the upstream face of the dam. This force adds to the dam weight, W, creating a total force, F, acting upon the center of mass (open circle). For the dam to be stable, we would like the total force vector to intersect the dam base within the middle third of the base.

vectors, which is the total force acting upon the center of mass of the dam. We would like this vector sum to be as near vertical as possible ($\theta = 0$) so that the total force pushes the dam into the bedrock foundation. This is not possible; the water force tries to shear the dam away from its foundation. Let us apply the rule of the middle third, mentioned in chapter 1 in the context of the stability of columns. We absolutely insist that the force vector meet the foundation inside the base of the dam because, otherwise, there will be significant shear forces acting upon the dam, but let us go further by adding a safety factor: the force vector meets the foundation within the middle third of the base. We can show that this requirement imposes a constraint upon the trapezoid angle φ: it must exceed 19°.

This minimum angle constraint is very similar to the result we obtained for the triangular dam cross section with our truss dam model. In both cases we have found that the forces which act upon a dam impose limitations upon the shape of the dam cross section. The mechanical properties of the dam material (the tensile and compressive strength of the concrete), the distribution of water force, the desire to reduce the uplift force—all these requirements lead us to a dam with an optimum base-to-height ratio. Different choices for dam cross section, and more detailed calculations, will lead to different results, of course. The main idea that I want you to take away from this analysis is that dam engineers have to do their sums before building their dam; for a given volume of concrete (that is to say, for a given construction cost) engineering analysis can point us toward the optimum shape for a dam.

SUMMARY: The four basic types of dam design (embankment, gravity, buttress, and arch) have been known since antiquity. Many large dams are hybrids. Embankment dams can be built from earth and rock fill; the other three types are usually constructed from concrete. Arch dams rely upon the strength of their construction material, whereas the other three types rely upon their weight to retain water. For the largest dams, the design choice is determined by the geology of the valley floor and walls. Large dams are massive constructions that influence the economy and physical topography of a region and require significant resources and know-how to build. We have seen from a simple truss dam model that the forces acting to overturn a gravity dam place constraints upon the dam's cross-section shape. The forces that act upon arch dams are more complex. There are many reasons for building large dams; in modern times the most important reason is often the huge amount of hydroelectric power that can be extracted from falling water.

CHAPTER 7

The Bigger They Are, the Harder They Fall

Moving Structures

Most of this book has been concerned with static physics. Our large structures are, we like to think, stable and stationary. One exception concerned rail bridges, which, you may recall, failed alarmingly in the nineteenth century because bridge designers did not at first understand that significant forces arise from the movement of heavy locomotives.[1] This chapter is all about architectural dynamics—in particular, about the collapse (intentional and unintentional) of tall buildings and smokestacks. In the first part of the chapter, I address engineering failure, but not all structures meet their demise through failures. Every structure has a finite lifetime; when a large building has outlived its usefulness or has become unsound due to natural aging, it must be brought down intentionally, in a safe and cost-effective manner.

The Leaning Tower of Pisa

We begin the chapter with an ancient and very elegant example of a collapsing building. To be sure, the collapse of the Leaning Tower of Pisa (fig. 7.1) is *adagio*—indeed, *lento*—meaning very slow, but it has been accelerating, until recently.

The tower at Pisa is the bell tower of the cathedral. Construction began in 1173 and lasted two centuries (there were many prolonged interruptions). During the extended construction process, the tower started to lean. Not alarmingly enough, it seems, for the builders to start over on a more suitable site with firmer foundations, but enough for them to con-

1. We encountered another exception where dynamics took over from statics when investigating the wind loading of suspension bridges.

FIGURE 7.1. A tower collapses—slowly. Thanks to Waldemar J. Poerner for this image of the Leaning Tower of Pisa.

struct the tower in such a way as to straighten it.[2] That is, the height between floors is slightly greater on the downhill side of the lean than on the uphill side. The tower leaned more and more as the seven floors were

2. The ground beneath the tower is unstable and unsuitable for supporting a heavy tower, especially since the building foundations were originally only 10 feet deep. The soil was formerly marshland and consists of sand, silt, and soft clay. The tower has sunk an average of 9 feet into this shifting base; the lean arose because the sinking is not uniform.

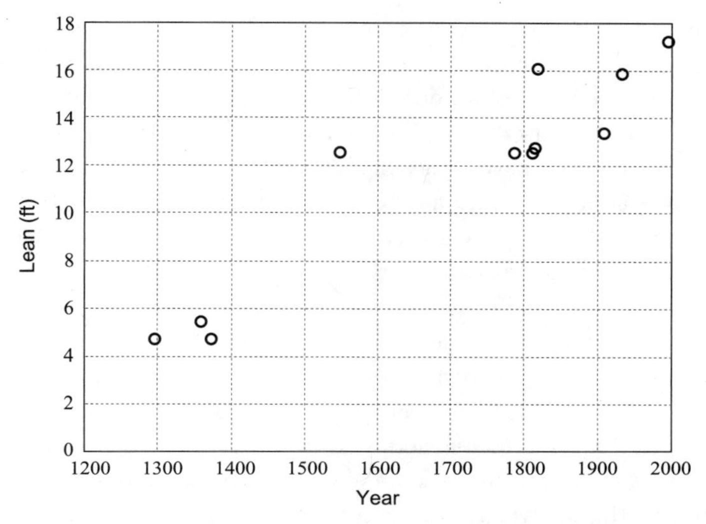

FIGURE 7.2. The progressive collapse of the Leaning Tower of Pisa, as recorded over the centuries by different observers. The scatter is due to the fact that different observers made their measurements in different ways, but the trend is clear.

built, and so the completed structure is not quite cylindrical, as it appears to be in figure 7.1, but is in fact banana-shaped.

The tower is about 200 feet high (more, before sinking) and 64 feet in diameter. It is built largely of marble and weighs 14,500 tons. On such unstable soil, this weight is unsupportable, and over the centuries the tower has sunk and leaned over to the south. Today, the lean amounts to about 5½°; the progression of the lean is shown in figure 7.2.

Many people have measured the inclination of the tower over the centuries, from different points and using different methods, so they arrived at different answers, as you can see from the scatter of the data points in figure 7.2 Today, sophisticated equipment is used to measure the inclination between the first and seventh *cornices* every 4 hours (plus measuring a host of meteorological parameters that have some influence upon the underlying soil).[3] So, modern measurements show much less scatter, which means that the rate at which the tower is falling over can be estimated quite

3. The cornices are the ornamental features that separate the different floors; you can see seven of them above and below the columns in fig. 7.1. The current lean of 5½° means that the seventh cornice protrudes almost 15 feet beyond the first cornice.

accurately. In the 1930s the Tower of Pisa was leaning at a rate of 4 seconds of arc a year ($4'' \text{ yr}^{-1}$); this accelerated to $6'' \text{ yr}^{-1}$ by the 1980s. We would expect the movement of a pencil standing on end to accelerate as it fell over—at a much faster rate, to be sure. Think of a pencil standing in an inch of molasses, and you have a reasonable model of the leaning tower of Pisa.

Over the centuries there has been much restoration of the tower. The inclination has been found to cause damage on the downhill side, especially near the bottom, because of the increased compression forces caused by the lean.[4] Eventually the lean would be so severe that, unless it is corrected, the building would collapse. Currently, the center of gravity is within the middle third of the tower, so that there is no danger of its toppling. After considerable thought about the best course of action, Italian engineers began remedial action in 1993. They added 600 tons of lead ingots to the north side of the foundation, thus compressing this side more than the south. The weight was later increased to 870 tons. The effect has been to stabilize the tower: it is no longer falling over. In fact, the peak incline of 5°33′36″ fell slightly to 5°32′51″ and has since held steady for several years. The authorities now have time to consider how to reverse the lean. The cost of fixing this problem, as you might expect, greatly exceeds the cost of preventing it. An unwise choice of a building site is less likely to be made today, at least in the developed world, thanks to building codes and an improved understanding of the physics of materials.

The beautiful Tower of Pisa has served its purpose here by introducing two notions that will recur in this chapter. First, the Leaning Tower provides an introductory example of human error in building construction. Second, it shows that modern building engineers can control the direction in which a building falls, given proper preparation. In the case of the tower, the fall has been stopped and will eventually be reversed; later, we will see how controlled building demolition can control the fall of other tall buildings—on a much faster timescale.

Towering Failure

We have just seen a historical example of the consequences of bad engineering. More recent, more rapid, and more tragic examples of building collapse

4. I modeled the forces that act on a tall tower in fig. 3.6 by constructing a truss tower, and indeed, as this figure shows, an asymmetric load induces large compressive forces near the tower base.

due to bad engineering practices abound, even today. Cheap apartment blocks will serve here to emphasize this point, though there are plenty of other building failures that qualify. In the middle of the twentieth century, cheap, affordable housing took the form of prefabricated tower blocks. Nowadays, such towers are still being built in less developed countries, with the same catalog of failures and loss of life due to either design or construction error. From news headlines over the past few years we find that a 12-story tower block collapsed in Alexandria, Egypt, in December 2007, leaving at least 30 people dead. This event was the latest of a string of such incidents in that country; unauthorized construction and building code violations have been mentioned in reports. In Morocco a half-built apartment block collapsed in January 2008, killing at least 16 people. Here, it was perhaps the case that concrete joists were not strong enough. Palencia, Spain, 2007: at least 3 people died and 25 were injured when a five-story apartment block collapsed following a gas explosion. Rome, Italy, 1998: at least seven dead when a five-story, 50-year-old apartment building collapsed due to structural failure—historical tunnels compromised the foundation.

Perhaps the most dramatic was the collapse of a 23-story prefabricated apartment block at Ronan Point in east London in May 1968. A gas explosion on a floor near the top destroyed a load-bearing wall, leading to the collapse of one entire corner of the structure. The building was new and had been occupied only a few weeks. Four people died, and 17 were injured. The collapsed wall placed an unsupportable burden on the apartment floor, which then collapsed onto the floor below, which in turn collapsed onto the floor below that, and so on. This domino effect of progressive collapse is often termed *pancaking*. I will analyze the pancaking effect in the next section of this chapter. The Ronan Point building was made of prefabricated concrete panels; the manner in which these panels were connected to each other, forming the building structure, was flawed.[5] Following its collapse the Ronan Point tower block was repaired

5. Buildings can be constructed according to a number of schemes. In situ concrete construction involves pouring a lot of concrete into molds and creating a single concrete structure, usually reinforced. Steel frame construction is based upon a skeleton of steel girders. Prefabricated construction is an inexpensive alternative that, in mid-twentieth-century Britain and elsewhere, gave rise (literally) to affordable high-rise housing. This LPS (large panel system) construction technique involved the casting of many concrete panels off site and then bolting them together on site. Clearly, for such a construction method the manner in which the panels are joined is critical to the building's strength.

and the building as a whole strengthened, after an inquiry that resulted in changes to the U.K. building code. Nevertheless, this collapse and others like it undermined public confidence in prefabricated structures. Eventually, the building was pulled down and replaced by low-rise housing.

In these examples of engineering failure, the reason for building collapse is immediately obvious (gas explosion), becomes apparent after investigation (poor foundations, faulty joists or connectors), or, we may reasonably assume, is due to shoddy workmanship (no or unenforced building codes, or unauthorized construction). In developed countries it is less common today to find any of these reasons for building collapse. In this book we are interested in large structures; they do not collapse for mundane reasons arising, for example, from unauthorized construction. The buildings of interest to us are high-prestige, high-cost, high-visibility, and just plain high. They collapse because a 180-ton aircraft with thousands of gallons of fuel is crashed into them, or because their reason for existence has gone away and so demolition experts carefully bring them down. Yet there are reasons for concern for all our large structures—not just buildings. They are designed and built by fallible humans. Construction techniques over the decades of the twentieth century have improved incrementally to an impressive degree. Building sites are safer for construction workers than they used to be, and the buildings, towers, bridges, and dams themselves are less likely to fail. Codes, along with manufacturing and assembly practices, include built-in checks and double checks. Nevertheless, there is room for concern. I am not suggesting that you should worry every time that you get into an elevator for the 80th floor or cross a long bridge; the concern I refer to is felt by engineers and designers before a structure is built.

Today, computers are an essential tool in the design and testing (by simulation) of large structures. A number of engineers are expressing concerns about our attitude to these useful machines and what they tell us. Increasingly we are relying on computers to do our engineering work; sometimes a design task is too complex for an engineer to evaluate by hand. The danger is that we are training a new generation of engineers who will substitute computer simulation for experimental testing, or who will believe a computer printout more than the practical experience of older generations. It is difficult to quantify this concern, which has as much to do with workplace ergonomics and education policy as it has to do with practical engineering, but it has been written about before now. Consulting some thoughtful essays on this subject will show you what I mean, in much more detail than I can provide here (see Petroski's *To Engineer Is Human* and the article by Ferguson listed in the bibliography). I will simply illustrate my point, and close this section, with a short story from my own experience as an engineer—one of several stories I could tell—of people placing too much reliance upon computers and not enough upon their own common sense.

I was discussing the testing of new military equipment with a Danish colleague in the 1980s. He told me of a test trial by the Danish navy of a new warship that had been equipped with the latest computerized navigation system. When the ship was approaching a harbor, the helmsman (and all the other naval personnel who were putting this vessel through her paces) peered intently at TV screens which showed them where the ship was located in a crowded harbor, as well as the harbor layout. It was claimed that the ship could maneuver in total darkness through a crowded environment without coming into contact with any other vessel and without the helmsman or anybody else assisting. In fact, the helmsman could in principle be safely below decks, since he required only the TV screens to navigate by. What happened is this: the navy engineers and sailors who conducted this vessel trial were all so intently looking at the TV screens that no one was looking through the windows, and the ship crashed into a small lighthouse at the harbor entrance. The lighthouse collapsed and fell onto the ship's deck. The red-faced navy personnel then had the embarrassing task of maneuvering their vessel (by eye, presumably) into harbor with a damaged prow and the remains of a lighthouse strewn across the deck.

Pancaking

The term *pancaking* is an informal yet expressive term that describes the manner in which certain tall buildings, such as apartment blocks, collapse progressively, as a top floor slams down on the floor below, causing this lower floor to collapse and so generating a structural landslide. Such pancaking collapses, when they occur, happen very quickly: the building falls almost as fast as if the top floor were in free fall, with no building structure beneath to slow down the fall. To the layperson, this result may appear surprising.

Consider figure 7.3, which illustrates an apartment block with an upper floor collapsing. In note 13 of the appendix I construct a simple physical model based upon Newton's laws that tells us how long it takes for this apartment block to collapse—that is, how long it takes for the floors to

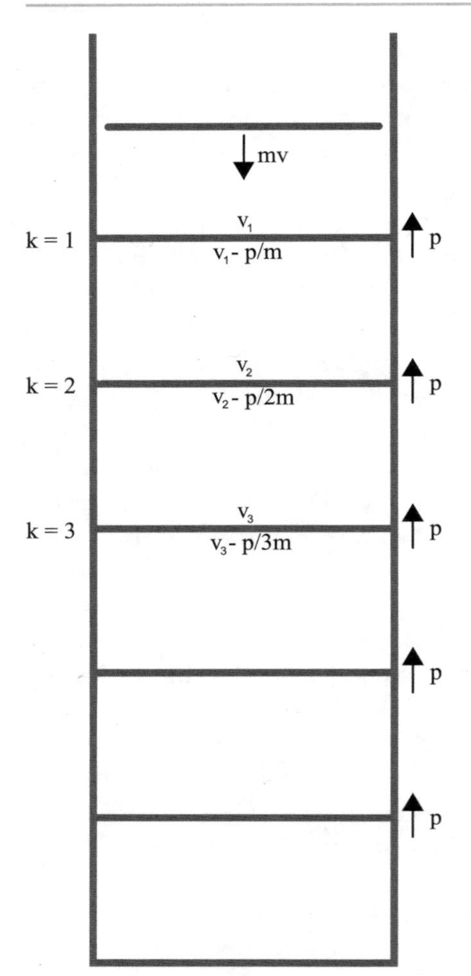

FIGURE 7.3. Pancaking. A floor collapses onto the floor below (labelled $k = 1$). The speed of the falling floor is reduced by the resistance of the $k = 1$ floor, but not enough to prevent both floors' falling onto the $k = 2$ floor. The speed of the falling floors is reduced by the $k = 2$ floor, but not enough to prevent all three floors' falling onto the $k = 3$ floor, and so on. The collapse accelerates as the mass of falling debris grows.

pancake down on top of each other and reach the ground. This model makes a number of simplifying assumptions that render the answer approximate, but it is pretty close to the truth. The result is shown in figure 7.4, where the collapse time (as a fraction of the free-fall collapse time) is plotted for buildings of different heights.

The collapse times of figure 7.4 are the *maximum* collapse times possible within my simple model: either no progressive collapse occurs (because the floors are strong enough to resist the impact from above), or the collapse is total and at least as fast as shown. In the next section I will mention an extension to this model that permits somewhat slower collapse times, but even with this extension the collapse time is not much greater than freefall time for tall buildings. The reason for this behavior is clear from the model. The mass of floors that slams down onto lower floors increases as the collapse progresses; the lower floors are less and less able to retard the downward progression of this mass as the collapse progresses. For tall buildings the mass is so great that the floor strength is insignificant by comparison with the force of falling debris; thus, the accumulated mass of floors effectively falls as if there were nothing in the way to impede it—free fall.

The tower block at Ronan Point and the other tower blocks that I mentioned in the preceding section collapsed because somebody made a mistake. These unintentional collapses led to pancaking; here, we have seen that pancaking buildings collapse very fast. Not all buildings collapse due to engineering error, of course. The two remaining sections of this chapter are devoted to the opposite extremes: big buildings that are collapsed intentionally. In the first case, terrorists intended and achieved great loss of life through uncontrolled destruction. In the second case, professional demolition engineers bring down large buildings safely, even in a crowded urban setting, via controlled explosions.

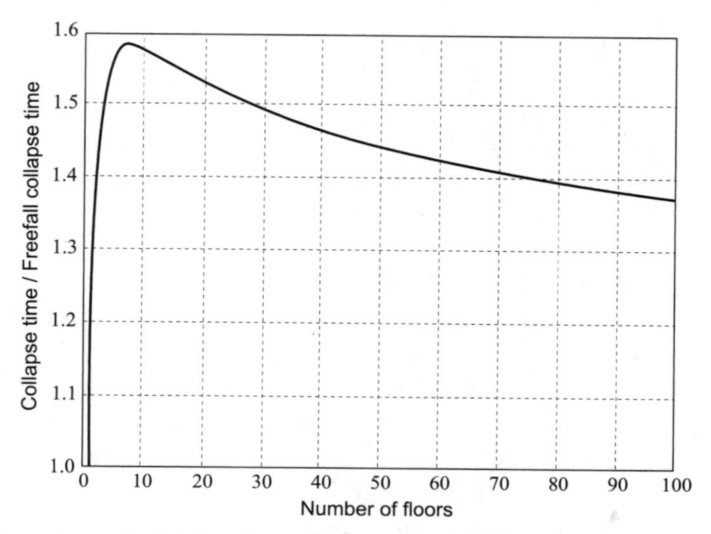

FIGURE 7.4. Apartment block collapse time (divided by freefall collapse time) versus number of collapsed floors. For tall tower blocks the collapse time approaches freefall time.

WTC 9/11

The title of this section would have seemed arcane to anyone who read the combination of letters and numbers before the World Trade Center tragedy in 2001. It is now, sadly, widely familiar. My purpose in including a section about these buildings' collapse is to illustrate some engineering lessons that have been learned and to inform you about construction engineering physics. This is the most difficult section of the book for me to write; I am acutely aware that some of my readers may have lost loved ones as a consequence of the terrorist attack. I have no intention of making any political points, of expounding conspiracy theories—in short, of ghoulishly raking over the ashes. Conspiracy theories? In researching this chapter I contacted a number of explosives demolition company representatives (such as Brent Blanchard of Implosionworld.com, who has kindly provided several images and answered a number of technical questions), all of whom, since 9/11, have to act very circumspectly with members of the public because of a few conspiracy theorists who would turn the tragedy of 9/11 into a farce by claiming that the WTC collapse was a controlled demolition. I most emphatically do *not* endorse this view and will give it no credence here. My discussion is restricted entirely to the engineering issues that arose from the collapse. Controlled demolitions will be discussed in the next section.

The Empire State Building easily survived a plane crash, as we saw in chapter 3. The 10-ton plane that crashed into the building in 1945 caused a fire that was extinguished within 35 minutes. The North Tower of the WTC survived a 1,500-pound bomb blast in an underground garage in 1993. So why did the plane crashes on 9/11 bring down both of the Twin Towers? A number of people have sought the answer in structural design faults, or conspiracies, but the balance of informed opinion is now quite clear: the buildings collapsed as a result of the airplanes' impact and the fires that these impacts generated. The energy and momentum of a 180-ton Boeing 747 moving at 470 mph is tremendous, and the impact damage caused to each building was severe. Recall, from chapter 3, that much of the strength of the WTC towers lay in their perimeter steel support columns. Many of these columns were fractured over many stories by the incoming planes, which then burst through and damaged the central core of each building. The initial crash destroyed the sprinkler system of each tower on the impact floors and ignited the planes' fuel, creating a fireball that led to high-temperature conflagrations (in places exceeding 1,100°C) engulfing several floors. The South Tower was extensively damaged from the 77th to the 85th floor, and the North Tower from the 93rd to the 99th floor. Both impacts also blew off fireproofing that clad the steel columns and truss floors, thus exposing the steel to the high temperatures generated by the fires. This last consequence of the crashes is the crucial factor; it is reckoned that both buildings would have remained standing had the fireproofing remained in place. (See the bibliography for the paper by Bažant et al., as well as the article by Eager and Musso.)

The design of the WTC towers included much redundancy as a safety factor, and so the loss of several perimeter columns and a simultaneous weakening of the central core were survivable. The worst building disaster in history resulted because the steel truss floors were heated up by the fires to such an extent that they were weakened and their mechanical strength reduced by half. As they weakened, the extra load on the floors immediately below each impact site increased due to debris falling from higher up. Each floor of the WTC towers was designed to support an extra 1,300 tons beyond its own weight, but the estimate is that 45,000 tons collapsed upon these floors due to impact damage and the resultant fires. Fifty-six minutes after impact the South Tower collapsed. Then the North Tower collapsed 102 minutes after impact. In both cases the buildings appeared to pancake; the heat-weakened floors that supported all the crash debris collapsed onto the floors below. Most likely it was the clips that connected the sagging truss floors to the perimeter columns that gave way. The floors fell rapidly, though not into their own footprints: much of the debris was thrown outward, as we saw in the dramatic photographs taken in the immediate aftermath (fig. 7.5).

The manner in which the North and South Towers fell has given rise to much speculation about conspiracies because the collapse looked superficially like intentionally engineered controlled collapses of tall buildings. In fact, a pancaking building does this simply through the force of gravity; as one commentator has said (overstating, somewhat), a half-million-ton building is not going to fall in any other direction than straight down. The speed of collapse also generated comments about controlled demolition. The North Tower collapsed only 60% slower than free fall; the South Tower collapsed only 40% slower than free fall—surely they could not have fallen so fast without assistance? In fact, as we have seen, pancaking collapses are very rapid. If you consult figure 7.4, you can see that the WTC collapses were rather slow for pancaking buildings. The reason for this was the strength of the two towers.

FIGURE 7.5. Ground Zero. Nearby buildings were extensively damaged as a result of the Twin Towers' collapse. U.S. Navy photo.

Recall that my pancake model of figure 7.4 assumed that the impact of just one floor would be sufficient to initiate a pancake collapse. The Twin Towers' floors were much stronger than the hypothetical tower block of appendix note 13, and consequently the collapse was slower. (The model of appendix note 13 can be extended to allow for stronger floors; in this case the model shows that the collapse time increases.) That the towers held up for 56 minutes and 102 minutes, respectively, permitted 99% of the people below the crash sites to evacuate the buildings. The perimeter steel design protected the buildings from collapsing upon impact; the structural strength of the buildings kept them standing as long as they did. The destruction of the WTC towers was not due to controlled demolition; we will see in the next section that such demolitions aim for pancake collapses —getting gravity to assist the fall straight downward. That so much debris was thrown outward, and that the collapse was relatively slow, tell us that it was not a controlled demolition.

The extensive investigations after the destruction of the World Trade Center towers led to a number of recommendations on tall building safety. Evacuation procedures can be improved, as can emergency communication and lighting. Fire protection of structural members can be made more robust, perhaps by encasement in concrete rather than spray-on asbestos.

It's a Blast

Controlled Demolition

The controlled demolition of large structures—usually tall buildings or blocks of buildings, or smokestacks—means the intentional destruction of the structure, such that it falls in a manner that is (as much as possible) predictable. We would like a smokestack to fall in a northwesterly direction, say, rather than southward because there is a building to the south that we want to keep intact. We try to make a skyscraper fall within its own footprint because there is no room for it to collapse in any other way without causing damage to other nearby structures.

Not all controlled demolitions require explosives, but many demolitions of large structures are carried out in this way. The explosives are placed at key locations, a process that may take many weeks. The magnitude of each explosion, and its timing, form part of an exquisitely orchestrated engineering symphony that aims to control the structural collapse as much as is humanly possible. Not only must the collapse be restricted spatially, but it must also be complete: a chimney or tall building that only half falls down is left in a dangerous state, and so further work—time-consuming, costly, and dangerous—must be carried out to complete the job.

The preparation may take weeks, and the cleanup may take months, but the demolition of a large building via explosive charges takes only seconds. A large dust cloud covers a wide area immediately after the *implosion*.[6] Consider the alternative: a comparatively slow demolition of buildings by more conventional means, at all times accompanied by the inconvenience

6. The term *implosion* has become standard for the type of controlled demolition via explosives that causes a building to fall into its footprint. From the physicist's point of view, this is an unfortunate word choice because *implosion* has another meaning: a sealed can that has the air pumped out of it implodes due to the force arising from atmospheric pressure.

to the public of a restricted worksite, continuous noise, and dust. For some demolition projects there is no choice—for instance, a building implosion might be too risky, or the explosives would spread toxic waste over an urban area. However, the logic for adopting a controlled explosive demolition is often compelling—it can save time and money—and so there are several large companies around the world that make a living by blowing things up (actually, blowing them down). Apart from demolition, these companies are sometimes consulted for forensic investigations to determine the cause of an unintentional building collapse (terrorist bomb or gas explosion?), and some are even employed by Hollywood to create spectacular special effects for movies.[7]

Consider a building that has been prepared for controlled demolition. The inside has been emptied, and the building stripped of all nonstructural material. Stiffening walls have been removed: they may retard the collapse.[8] Critical supports have been identified, and hundreds of different charges have been put in place, connected by miles of electrical cable. Blasters have drilled holes into selected concrete columns and pounded just the right amount of dynamite into the holes (modern dynamite is so stable that it can be safely pounded). A specialized high explosive, RDX, has been formed into linear-shaped charges that will slice through steel columns and paired with a separate dynamite charge that will kick aside the remnant sections. The columns to be blasted are covered with chain link fencing and heavy geotextile fabric to reduce flying debris; a curtain of fabric also covers the windows of each floor that is to be blasted. Detailed computer simulations have been used, along with the blasters' experience, to decide where the charges are placed and how the sequence of detonations is orchestrated. Sometimes a few test blasts have been carried out ahead of time to provide the blaster with a more precise idea of how much explosive material is required for the support structures of the building. A miscalculation can be costly: too few explosives and the building might

7. I suppose that such movies would be "blockbusters."

8. The lower floors of the WTC towers had not been weakened, so the building was strong enough to slow down the eventual collapse, and the lower floors had enough material in them to prevent a purely downward collapse—much of the debris was thrown outward. The initial explosions were high up; controlled demolitions place the initial explosions near the ground to facilitate pancaking. These basic reasons tell demolition experts that the WTC disaster did not result from controlled demolition. For an expert's informed opinion, see the article by Brent Blanchard in the bibliography.

FIGURE 7.6. Controlled demolition of an apartment block in a crowded neighborhood. Note how the building is falling into its footprint. Thanks to Brent Blanchard for this image.

hang up—remain partially standing; too many and debris may fly far and wide, causing damage to other buildings or killing onlookers a quarter mile away. A portable seismograph has been set up to measure the ground blast, and video cameras will record the collapse from different angles. The information contained in these recordings will add to the blasters' experience, for future reference. Now the blasters retire to the detonator controls and commence countdown: sirens sound at 10 minutes, 5 minutes, 1 minute.

A blaster presses and holds down the "charge" button on a detonator; when a light comes on to indicate that enough of an electric charge has been built up to set off the blasting caps, he (or she) presses the "fire" button. About one second after this initiation of the detonation sequence, the first of many hundreds of explosions occurs at one end of the ground floor. More explosions ripple across the ground floor and other sequences begin, higher up (see fig. 7.6). After about 10 seconds the first external signs of collapse: one end of the building begins to fall, exerting a pull on a side wall and dragging it inward and down. After 12 or 15 seconds the ripple of charges has reached the far end of the building, cracking the last of the support columns. A few seconds later, there is an outward rush of air expelled from the lower floors as upper floors concertina, pushing out a vast cloud of dust. Then comes the cheering. After 15 minutes or so the dust clears, and the postmortem and cleanup can begin.

This building might have been part of a city center, surrounded by other buildings that were not to be damaged—indeed, that may still be occupied. Thus, it is important that the building fall into a very restricted space. These implosions are the most difficult projects undertaken by the demolition companies, and they are often stunning (metaphorically speaking) public spectacles. How has the industry reached such a level of knowledge and experience that it can reliably and safely carry out the destruction of a structure that weighs tens of thousands of tons, using hundreds of pounds of high explosives? The history of controlled demolitions really began a century ago.

Implosion Evolution

Sappers and military engineers have been using explosives to destroy enemy buildings and bridges for hundreds of years, but of course military practice is very different from controlled demolition. An eighteenth-century sapper planting explosives to bring down the wall of an enemy fortification would not have been concerned about public safety; indeed, he may have intended to inflict great loss of life. A World War II military engineer blowing up an enemy bridge would not have been concerned about using too much explosives or about how far the debris flies. The use of explosives in peacetime to bring down structures that are no longer wanted began in the early twentieth century and was initially a haphazard and often unsuccessful enterprise. Trial-and-error methods based upon military experience and upon the expertise of rock-blasters were applied to isolated buildings with indifferent results. The very idea of using dynamite for the purpose of building demolition was new—encouraged by dynamite manufacturers who saw a new means of advertising and of promoting sales.

In these early years there was little concern about the effects of explosive demolition outside the worksite. Protection was minimal. There were no delay devices, just one large blast, and so bigger buildings required bigger charges of dynamite. Consequently, each explosion generated significant flying debris, a loud air blast, and a violent ground vibration. By the 1930s the art of controlled demolition (it was not yet a science) saw the introduction of protective measures, which were applied in built-up areas to restrict flying debris. Windows were covered by wooden boards; demolition company trucks were placed around the blast zone perimeter; dynamite-packed bore holes were covered up. Explosive demolition of structures was becoming more predictable and professional by the end of this decade.

Europeans gained a lot of experience in controlled demolition by clearing up bomb sites in devastated cities across the continent in the years following World War II. Indeed, in Britain a particular position, the "special commissioner of clearance," was set up for this purpose. Also during these immediate postwar years, staggered small blasts were introduced, a technique borrowed from rock-blasting. This technique gave the blasters more control over the building collapse and also minimized the ground vibration and air blast. Great advances in the theory and practice of controlled demolition were made during this period because there was so much clearance work to do and so many buildings to practice on.

By the 1960s techniques were more refined. Steel beams that formed part of the structure of a building that was to be demolished were precut. Cables were attached to the building at key points to pull the collapsing structure in the desired direction. The high explosive cyclotrimethylenetrinitramine (better known as RDX) was developed. RDX in shaped charges directs a thin, high-velocity blast that slices through steel. Nonelectrical firing systems appeared; nonelectrical delays were a significant development for blasters because these made demolition safer and the blasting sequence more reliable. Controlled demolition first caught the public eye in the sixties because demolition teams were by this time sufficiently expert to demolish structures that were within spitting distance of occupied urban settings. The 1970s saw TV airing of controlled demolitions. This decade also saw the introduction of new protective measures such as the use of geotextile fabrics around building columns and the placement of large steel containers around blast site perimeters.

In the 1980s the art was becoming a science. For both safety and economy it was desirable to bring down a building using the minimum possible amount of explosives. By now it had become practicable to calculate this minimum charge, based upon experience and a few test shots. The following decade brought lucrative contracts to demolition companies as the Cold War ended and the military needed assistance in destroying missile sites and superfluous weapons. At the same time those prefabricated 1960s apartment blocks were reaching the end of their days.

I mentioned earlier that 9/11 has had an impact on today's controlled demolition industry as a consequence of the misguided statements from some conspiracy theorists. There were other influences of 9/11: in the immediate aftermath, demolition projects were suspended across North America and Europe because of security concerns; in addition, regulations covering the storage and transportation of explosives were tightened. Today the main issue facing controlled demolition companies is not the technical challenge—it is the mountain of paperwork that they must face.

Mistakes are made even now, but fewer and fewer as blasters learn from experience.[9] In 1997 a young spectator was killed in Australia as she stood watching a demolition, 500 yards from the blast site, when large chunks of debris were thrown beyond the site perimeter. Buildings "hang up" because of bad luck or because of miscalculation. *Cutoff* can interrupt a sequence of charges resulting in only partial collapse. Cutoff occurs when a line is severed by an explosion, thus cutting off the firing sequence further down the line. Generally speaking, however, reliability and predictability have increased greatly from the early days, so much so that controlled demolition is now routine in congested urban sites. You can see many videos of controlled demolitions on the Internet (see, for example, the Nantais and ImplosionWorld entries in the bibliography).

To give you an idea of the scale of some of these implosion or controlled demolition projects, here is a list of the largest structures that have been brought down by explosives.

— *Largest*: Sears Merchandise Center, Philadelphia. Twelve thousand pounds of explosives were used to reduce 2.7 million square feet of building to rubble in 1994.
— *Tallest building*: Hudson's Department Store, Detroit. This 26-story, 439-foot-tall building was imploded in 1998. (See Nantais in the bibliography.)
— *Tallest free-standing structure*: Matla Nuclear Power Station, Johannesburg, South Africa. A 906-foot concrete smokestack at this power station was demolished via controlled explosions in 1982.

9. Even today there is no school for teaching controlled demolition; each blaster learns on the job. One of the reasons for this is that every job is different. Even if a number of identical buildings are to be brought down, each is assessed and prepared individually.

— *Longest*: The Ohio Turnpike Bridge, Interstate 80 near Boston Mills, Ohio. This steel truss bridge, which was 2,680 feet long, was demolished in 2003 with 138 pounds of RDX.

Demolition Physics

The basics of controlled demolition physics are shown in figure 7.7. If we want a tower or smokestack to fall eastward, then we knock out the east part of the tower base, as shown. I hope that this revelation has not stretched your neurons too much—the physics does get more complicated. Now consider the prefabricated apartment block shown in the figure. We need a building such as this one to collapse downward into its footprint and not fall to the side, where it may damage nearby buildings. If we knock out the ground-level column shown, the weight it supported brings down one end of the building while at the same time pulling in the right side wall. A delayed demolition of the second support column shown in the figure will cause the left part of the building to fall down and in. Why delay the second explosion? Well, it could be done simultaneously and result in the same desirable outcome, but the ground vibration would be more severe. If the second explosion is delayed, the ground vibration is twice as long but only half as bad. This approach—of using many small charges rather than one enormous charge—has been likened to pouring a sack full of sand onto the ground instead of dropping the whole sack all at once. Thud.

What if the apartment building is strengthened with roof trusses, as shown in figure 7.7e? You can see (now that you are an experienced expert at truss analysis) that the original approach won't work. Knocking out one or even two of the base columns may not be enough to bring the building down. Even if it does, the roof stiffening may cause the building to fall awkwardly, instead of collapsing like a house of cards as it did before. For this new situation you will have to precut the extra roof truss members or destroy them with explosives, so that they cannot resist the inward collapse of the walls. What if the apartment building has a central core of reinforced concrete, as many of our skyscrapers do? You will have to demolish the core all the way up, as suggested in figure 7.7f, if the structure is to collapse totally.

The simpler case of dropping smokestacks in a chosen direction (fig. 7.7a–b) contains some interesting physics, though this is of little interest to demolition blasters. You may have seen photos or video footage of old-

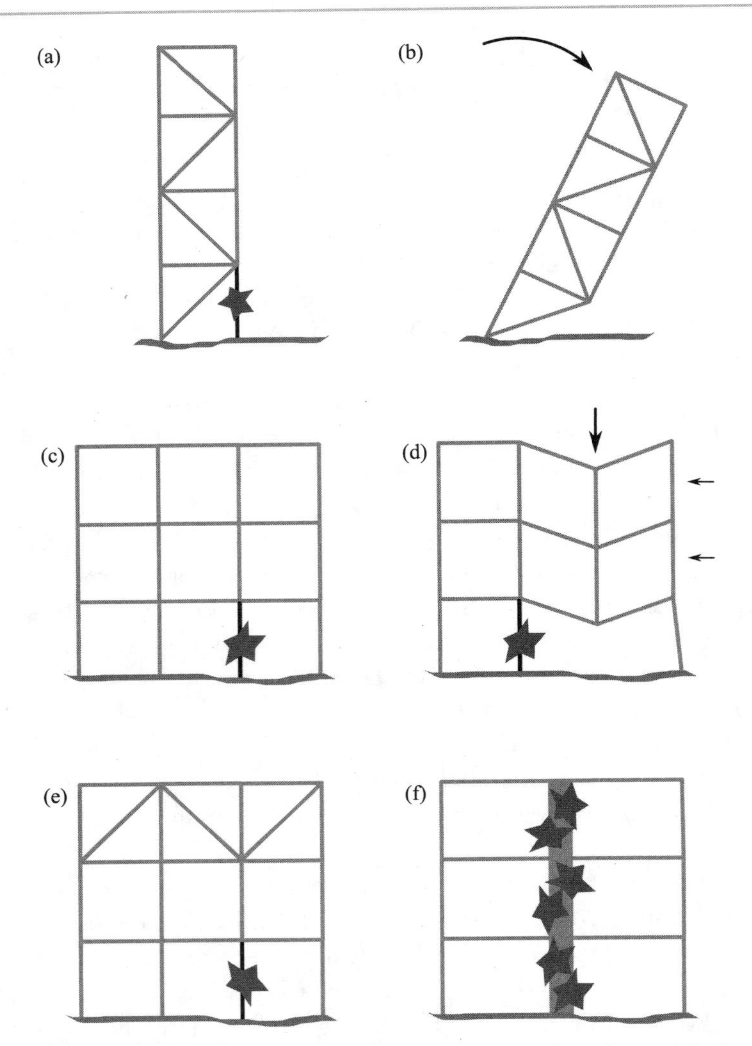

FIGURE 7.7. Controlled demolition 101. Knocking out a supporting column at the base of a tower (a) causes it to fall in a chosen direction (b). Knocking out this support column of a prefabricated apartment block (c) causes upper floors to fall and pulls in the right side wall (d). (e) The same explosion is not enough to bring down this building. (f) For buildings with a central core it will be necessary to destroy the whole core.

fashioned tall brick chimneys and noticed that the structure breaks in the middle *before* it hits the ground. That is to say, the initial explosion at the base causes our chimney to fall in one piece, but before it hits the ground, the chimney then breaks in two. Why? You may recall, way back in chapter 1, that two forces can cause a structure to break in this way: bending force and shear force. We have not had to deal with these beasts much since chapter 1 because of my truss analysis approach, which reduces complicated problems involving all kinds of forces to simpler problems requiring only compression and tension forces. A detailed analysis of a falling brick chimney shows that it breaks usually in one of two places: at the base or about a third of the way up from the base. Not all chimneys break, because some of them (metal or reinforced concrete chimneys, for example) are strong enough to withstand the bending and shear forces as the structure topples. Brick structures are not strongly resistant to these forces, however. When they break, they always do so at the front edge, so

FIGURE 7.8. A masonry chimney usually cracks about one-third of the distance up, when it falls. Thanks again to Brent Blanchard for this image.

that the upper part of the falling structure is more upright, as shown in figure 7.8.

Analysis shows that the shear force and the bending force both vary along the length of the chimney, and they both increase as the fall angle increases. At any given angle the shear force is at a maximum at the chimney base, while the bending force is at a maximum at a point one-third of the distance up from the base. This is why the brick chimney breaks at one of these places.[10]

SUMMARY: How large buildings collapse—whether that collapse is intentional or a catastrophic accident—depends upon their construction. Tall apartment blocks pancake, falling almost as fast as if they were in freefall with no internal structures to slow them down. Controlled demolition can cause such a building to collapse into its own footprint, with little debris outside this area. The methods used for achieving controlled demolition—strategically placed high explosive charges plus weakened support structures—are very different from the methods used by terrorists to bring down the WTC towers, and the resulting collapses are different. Demolition expertise has been built up over the past century and is based upon practical experience coupled with some technical advances. The manner in which simple structures such as brick smokestacks collapse is understood in detail.

10. See the Varieschi and Kamiya article listed in the bibliography, and also my paper. Varieschi's website has many examples of falling chimneys and toppling brick towers that illustrate the relevant physical principles.

AFTERWORD

Highbrow Engineering, Heavyweight Art

Architecture is a cross-disciplinary subject, with one foot firmly planted in the realm of science and engineering and the other in the realm of art. I do not mean to imply by this statement that these two realms are wholly unrelated: they are connected by the bridge (an elegant arch bridge, I fancy) of mathematics. Many mathematical constructions are considered to be art—fractals, for example—and many artistic themes such as symmetry are reflected (now that is almost a pun) in mathematics.

In this book I have explored the technical, functional side of our largest structures through the lens of physics, and physics is the most mathematical of sciences. Look at how much geometry we have considered in these chapters: straight edges joined to make triangles, rectangles, cubes, tetrahedra, and trapezoids. Curves in two or three dimensions, such as catenaries, circles and ellipses, cylinders, cones, pendentives, hemispheres, hyperboloids, and hyperbolic paraboloids.

Put together the notions expressed in the two preceding paragraphs, and you will see where I am going with this final, short essay. Many people consider that architectural constructions—those that require a significant mathematical input—may exhibit significant artistic merit. Indeed, some of these structures can be considered stand-alone art exhibits (the Eiffel Tower, perhaps, or the Taj Mahal). The architects, engineers, and builders who labored to make the large structures we have investigated may have had aesthetic appeal in mind from the outset, of course; and the artisans and craftsmen who adorned some of these structures certainly were appealing to our artistic appreciation. My claim here is that even without intentional adornment, many of our largest engineering structures possess aesthetic appeal because of the mathematics that they exhibit, be it geometry or symmetry or whatever. There is a majesty of scale that quickens the pulse, or a sweeping curve that draws the eye. This claim, I am sure, is neither original nor surprising, but in the presence of a monumental piece

FIGURE A.1. Decorated fan vaulting, Paris. Thanks to Don Cooper for this image.

of structural engineering (such as the Hagia Sophia or the Gateway Arch), I have always found it compelling.

First and easiest, we can see art *on* some large structures. The walls, ceilings, columns, and windows of medieval European churches were luxuriantly adorned with artistic motifs. Figure A.1 is an example of this type of architectural art. Other examples in this book include the French courtroom ceiling seen in figure I.3, and any of the decorated dome interiors we have examined. Only a little less obviously, we see artistic or at least aesthetic merit in some structures that may exhibit only straight lines, such as a skyscraper or a vaulted ceiling. The fact that not all skyscrapers and

FIGURE A.2. Exuberant scaffolding: richly sculpted flying buttresses in Milan, Italy. I am grateful to Glenn Sanchez for this image.

ceilings appeal to us shows that some design effort beyond the purely functional must have been made by the builders or architects of those that do appeal.

We have seen how even the purely functional engineering facets of a structure can be made pleasing and visually appealing. The fan vaulting of Bath Abbey (fig. I.2) is a striking example, as is the massive and wholly unapologetic scissor buttress inside Wells Cathedral (fig. 3.5). Buttresses are a particularly good example of engineering art. They are functionally very mundane—nothing more exciting than scaffolding—but because they are part of high-status structures and because (by their very nature) they cannot be hidden, much effort has been poured into ensuring that they look good. Note how the flying buttresses of figure A.2 become part of the structure, like the Wells scissor buttress.

"Engineering art" is also visible in half-timbered houses. Once again there is no attempt to hide the structural aspects of the building; on the contrary, the beams are exposed because they are beautiful. The trusses that constitute the Eiffel Tower or many of our large suspension bridges are exhibited instead of hidden; like dinosaur bones they impress us with their scale, strength, and proportion. We have seen humble brickwork made into appealing geometrical patterns that are functional. We have seen sweeping curves that were chosen or that arise naturally (the Hoover Dam arch or the Golden Gate suspension cables, for example).

Big structures impress because of their height or depth; there is a conflict between their simple shape (say, a hemispherical dome) and their scale that fascinates us. Part of this fascination, for me, is the juxtaposition of scale. We are not used to seeing simple shapes—very straight lines or smooth curves—that are very large because the natural world offers few of them. The curved horizon seen from an airplane is one exception: precisely because it is unusual, it holds the gaze of many a traveler. At least part of the conflict that is set up in our mind's eye when we see a large engineering structure is between the simple shape (think small) and the observed depth and perspective (think large).

An appreciation of the beauty of large buildings, dams, and bridges is icing on the cake. We are already predisposed to be fascinated by these largest examples of human creativity because we can appreciate the effort and organization that went into their construction. I believe that my fascination with big structures—both scientific and artistic—has been magnified by an understanding of the physics and mathematics that went into their construction. I hope that this book enables you to look anew at our largest structures with enhanced appreciation.

Technical Appendix

Note 1: "Simple" Cantilever Beam Theory

The results of a beam theory calculation show that the deflection, y, of the end of the beam shown in figure 1.15 is

$$y(L) = -\frac{L^3}{EI}\left(\frac{1}{3}W + \frac{1}{8}w\right) - \frac{L}{nA}\left(W + \frac{1}{2}w\right). \tag{TA1.1}$$

In this equation w is the beam weight, W is the external vertical load applied to the end of the beam, L is beam length, and A is beam cross-sectional area. The parameters E, I, and n require some explanation. E is Young's modulus, introduced in chapter 1. This parameter is a measure of the resistance of a material to elastic distortion: larger E means less deformation of a material for a given load. Young's modulus for steel is about twice that for cast iron, and about 10 times that for concrete, so a steel beam deflects less, under a given load, than a cast iron beam of the same shape, and much less than a concrete beam of the same shape. I is a parameter much loved by structural engineers, but dull as ditchwater to most other human beings. It is the *area moment of inertia* and is a property of the beam shape: it has nothing to do with what the beam is made of. A large I means that the shape is resistant to bending; a small I means that it will bend easily. The parameter I can be calculated from beam theory for any beam and is a function of the beam dimensions only. The product EI occurs often enough to be given its own name: *flexural stiffness*. The parameter n of equation (TA1.1) is the *shear modulus*; it describes the resistance of a material to shearing forces. A material with large shear modulus will not shear easily.

From the foregoing we see that the first term of equation (TA1.1) tells us how much the beam end is deflected due to bending, while the second term tells us the deflection due to shear. So, a real beam subjected to a load can both shear and bend. The bending contribution depends upon the cube of beam length, whereas the shear term is only linearly dependent upon length. This shows that bending is more important for long, thin beams, whereas shear is more important for short, thick beams. The form of equation (TA1.1) is not obvious—we could not have guessed that the beam deflection would behave as described by this equation without a detailed beam theory calculation. Recalling that this describes a *simple* cantilever beam, you can perhaps now readily appreciate that beam theory applied to more complicated structures is going to be pretty horrendous, which is why we will not go there.

Applying beam theory to the lintels of Stonehenge, or to the beam shown in figure 1.9a, we find that the beam will support a load that increases with Young's modulus, E, and with beam width, as we would expect, and that decreases with the fourth power of beam length, L^4. More surprising, perhaps, the load our beam can support increases as the *square* of beam depth. So length and depth are more important that width or cross-sectional area. This fact is related to the tension and compression forces that act upon the top and bottom surfaces of the beam, and explains why ancient stone lintels are so short and thick (fig. 1.1).

Note 2: Buckling Formula

The load that is required to cause a column to buckle is predicted by the Euler column formula, which is

$$F = \frac{\pi^2 EI}{(kL)^2} . \tag{TA2.1}$$

Here $\pi = 3.14159$. The factor k depends upon the manner in which the top and bottom ends of the column are fixed. If both ends are hinged, then $k = 1$; if both ends are fixed firmly, then $k = ½$; if one end is free to move laterally, then $k = 2$. For different combinations (for example, if one end is fixed while the other is pinned) k can assume other values, but in all cases k is close to 1. Euler's formula tells us that the force required to buckle a thin column increases with the column's flexural stiffness (no surprise there) and decreases with the column length (no surprise there, either). The buckling force is quite sensitive to length: if you triple the length of a column, the buckling force falls by a factor of 9. Again we see that the strength of a structural component depends upon both the component material and upon its shape.

Buckling can occur as a result of thermal expansion. We can calculate the amount of buckle if we know how much a material expands with heat. The maximum buckle of a steel roof of length L is

$$h = \frac{1}{4} L\sqrt{6\delta} . \tag{TA2.2}$$

Here δ is the fraction of length that the roof expands for a given rise in temperature. Let us say that an inexperienced do-it-yourself roofer has nailed a sheet of galvanized steel to a roof in midwinter, when the temperature was $-20°C$ ($-4°F$), and he did not allow for thermal expansion—say he placed a line of nails along the ridge of the roof and another line along the lower edge. The two lines are 10 feet apart. The following summer the temperature increases to, say, 20°C (68°F). Steel expands at the rate of 0.000012 inches per inch of length per degree centigrade increase in temperature. In our case, therefore, the fractional increase in roof length due to thermal expansion is $\delta = 0.00048$, and so from equation (TA2.2) the steel sheet will bulge about 1.6 inches. Of course, professionals allow for expansion when fixing metal roofs, just as rail builders separate rail tracks with expansion joints.

Note 3: Elementary Truss Calculation

In this note I provide the calculations for the triangular truss structure shown in figure 2.4, to find the tension and compression forces that act along the length of the three members. Consider the pin on the left. The total force that acts upon this pin must be zero. So, taking first the horizontal components, we find that

$$T_2 + T_1 \cos(a) - X = 0. \qquad \text{(TA3.1)}$$

We assume that the stress within each member is a tension (if it turns out negative, then it is a compression), and so $T_{1,2,3}$ point *away* from the pins. The vertical components yield

$$Y_1 + T_1 \sin(a) = 0. \qquad \text{(TA3.2)}$$

Similarly for the pin on the right:

$$-T_2 - T_3 \cos(b) = 0, \qquad \text{(TA3.3)}$$

$$Y_2 + T_3 \sin(b) = 0. \qquad \text{(TA3.4)}$$

The apex pin gives us the last two equations:

$$T_3 \cos(b) - T_1 \cos(a) = 0, \qquad \text{(TA3.5)}$$

$$-W - T_3 \sin(b) - T_1 \sin(a) = 0. \qquad \text{(TA3.6)}$$

Solving these six equations we find, for the six forces,

$$X = 0,\ Y_1 = W\frac{\sin(a)\cos(b)}{\sin(a+b)},\ Y_2 = W\frac{\sin(b)\cos(a)}{\sin(a+b)}, \qquad \text{(TA3.7)}$$

$$T_1 = -W\frac{\cos(b)}{\sin(a+b)},\ T_2 = W\frac{\cos(a)\cos(b)}{\sin(a+b)},\ T_3 = -W\frac{\cos(a)}{\sin(a+b)}. \qquad \text{(TA3.8)}$$

That the ground exerts zero horizontal force ($X = 0$) is not a surprise because there is no external horizontal force applied and the truss is not moving. Similarly, we expect that the load, W, will equal the upward forces, and indeed $W = Y_1 + Y_2$. Both upper members are under a compressive stress because the load, W, is pressing down. The horizontal member is subjected to a tensile force, which makes sense if you think about it. Substituting the angles of figure 2.4 ($a = 25°$, $b = 60°$) we obtain the magnitudes of the forces shown in figure 2.4. Truss analysis has enabled us to solve completely for the forces that act within the members of this truss.

Note 4: Buttresses

For the buttress on the right in figure 3.2b to be effective, the total force vector (the vector sum of vertical and horizontal loads) must be directed through the buttress base and not out the side. If the force is directed through the base, the load is taken up by the ground beneath, and the building is stable. If, on the other hand, the load is directed through the buttress wall, the wall may shear; tensile forces are set up which the wall and buttress may not be able to resist. The condition that must be satisfied by the buttress in order to resist the horizontal forces is $d_1 > h_1 / \tan\theta$, as we can show from the figure by simple trigonometry. Here, θ is the roof angle, h_1 is the height of the wall and buttress, and d_1 is the buttress width.

On the left of figure 3.2b we have a flying buttress. The key to understanding flying buttresses is to observe that the column is tall to supply extra weight, the purpose of which is to push the total force vector downwards, making it more nearly vertical. This occurs because the downward vector component is increased by the buttress weight, whereas the horizontal vector component is unchanged by this weight. So, stability is achieved with a narrower buttress. It is not difficult to show from the geometry that the required flying buttress width must satisfy

$$d_2 = \left(h_1 - h_2 \frac{B}{B + \frac{1}{2} W} \right) / \tan\theta , \qquad \text{(TA4.1)}$$

where h_2 is the flying buttress height shown in figure 3.2(b) and B is the buttress weight. You can see that d_2 is less than d_1.

Note 5: Natural Frequency, Damping, and Resonance

Imagine a pendulum that consists of a metal rod attached to a pivot at the top with a weight at the bottom. This pendulum will oscillate with a frequency that depends upon the length of the rod. The frequency is pretty much the same for all rods of the same length, and for all weights, and so we call it the *natural frequency*. Now watch the pendulum oscillations die down to nothing owing to air resistance: this is *damping*. The damping would be much greater if the pendulum were oscillating underwater. Damping sucks energy from the pendulum.

Now imagine energy being put back into the pendulum by an electric motor that causes the pivot to oscillate back and forth, at a frequency we can choose. How does the pendulum react to this new source of energy? It oscillates at the motor frequency, a phenomenon known as *forced oscillation* to many generations of physics students. The amplitude of the forced oscillation—how big the pendulum swing is—increases greatly when the motor frequency is close to the natural frequency. This increase in oscillation amplitude is known as *resonance*. Many oscillating physical systems exhibit these effects. Systems that are much more complicated than a simple pendulum (such as spin driers, lathes, or other rotating machinery) may exhibit more than one natural frequency and more than one resonant frequency.

A familiar example of resonant oscillation occurs in your domestic tumble drier or washing machine. You may have noticed that the drier drum, as it slows down at the end of a spin cycle, briefly undergoes high-amplitude vibrations. These vibrations occur when the spin rate passes through the drum resonant frequency.

Note 6: Catenary Equation

The catenary equation is

$$F(x) = \lambda \cosh(x/\lambda) = ½\,\lambda(\exp(x/\lambda) + \exp(-x/\lambda)). \qquad \text{(TA6.1)}$$

Imagine drawing a horizontal line below the lowest point of the chain. In equation (TA6.1), x is a distance along this line, and $F(x)$ is the height of the chain above the line at the point x. The parameter λ (Greek *lambda*) describes how flat or curved the catenary function is; many different catenary curves exist, each corresponding to a unique value of λ (for the necklace of fig. 4.3, λ is approximately 2¼ inches). So, in short, equation (TA6.1) describes a family of curves, and one equation of this family describes the shape of the necklace in figure 4.3, while others describe a heavy anchor chain or a taut rope or whatever. Make the chain more taut and the catenary has a larger value of λ; slacken the chain so that the curve is more pronounced, and the corresponding catenary has a smaller value of λ. Physically, the parameter λ is the ratio T/w, where T is the horizontal component of the tension force and w is the chain weight per unit length.

The proof that chain curves are closely approximated by catenaries can be made in a couple of ways: we can either consider the forces that act upon the chain, or we can assume that the chain adopts the shape that minimizes its energy. Both methods lead to the catenary. Both methods also involve calculus. The derivation can be found in many math, physics, or structural engineering textbooks, and also online: just search for "catenary."

Note 7: Truss Arch Analysis

Figure 4.9 shows a five-element truss that approximates an arch. The truss shape is defined by two angles, θ and φ, that specify the depth or thickness of the arch as well as its flatness. Two horizontal loads, F, are added. These are not necessary for the structure's stability but will teach us something about how buttressing influences the nature of the forces that act within the arch. Algebra and perspiration on the part of your hard-working author determines the three forces $T_{1,2,3}$ of figure 4.9 to be given by

$$T_1 = \frac{1}{2\sin\varphi}\left[\,2F\sin\theta - W\cos\theta\,\right], \qquad \text{(TA7.1)}$$

$$T_2 = \frac{\sin(\theta+\varphi)}{\sin\varphi}\left[\,W\cos\theta - 2F\sin\theta\,\right] - \frac{1}{2}W, \qquad \text{(TA7.2)}$$

$$T_3 = \frac{1}{2\sin\varphi}\left[\,W\cos(\theta+\varphi) - 2F\sin(\theta+\varphi)\,\right]. \qquad \text{(TA7.3)}$$

In words: the forces from external loading that act within the arch are greater for thin arches (small φ) than for thick arches. Without external buttressing (i.e., if $F = 0$) a compressive force acts on the top surface of the arch, while a tensile force acts on its undersurface. For the architectural masonry arches of olden times this is no good—the stone cannot be in tension—and this is why they need external buttressing. We can show from equations (TA7.1)–(TA7.3) that the magnitude of the buttress force F is constrained by the need to keep the members of the truss arch in compression. For example, if we choose $\theta = 30°$ and $\varphi = 5°$, then F must exceed $0.714W$ but be less than $0.866W$ to keep the upper and lower surface of the arch truss in compression. Also, F must exceed $0.790W$ if the force T_2 is to be in compression. This force represents what is going on inside the arch structure. Therefore, buttressing forces must be set up carefully for an arch constructed of material that is weak in tension, so that only compressive forces act on every part of the arch.

Note 8: Equation for a Parabola

A parabolic curve can be written as follows:

$$F(x) = ax^2. \tag{TA8.1}$$

Here $F(x)$ is height above the lowest point of the curve, and x is horizontal distance from this point. The parameter a tells us how steeply the curve rises. A bridge arch is described by an inverted parabola, such as $F(x) = H - ax^2$.

Note 9: Suspension Bridge Span

What is the theoretical limit for suspension bridge span? Well, we can calculate the maximum length for a steel cable and cut down from that figure. A steel cable suspended from one end has a maximum length of $L = S/\rho g$, where S is the yield strength of cable steel (see chap. 1), ρ is steel density, and g is the acceleration due to gravity. If the length exceeds this value, the cable will break under its own weight. Plugging in numbers we find $L = 13$ miles. Because cables are suspended at an angle on a bridge, rather than being hung vertically, the maximum bridge length is actually less than this value. The cables must support the bridge deck weight as well as their own weight, and that reduces it more. Add in a safety factor, and we come up with a realistic value for the maximum suspension bridge span of about 2–3 miles.

Note 10: Loosening the Bolts on Lion's Gate Bridge

Consider figure 5.21. The bridge deck is at the apex of the triangle, and here is where the mass of the bridge is concentrated. Let us imagine that an earthquake shock shifts the ground upon which the piers rest, imparting a lateral force, F, to the right. This is equivalent, for my purposes here, to saying that a horizontal force to the left of magnitude F is applied to the apex of the triangle.[1] Truss analysis leads us to the following expression for the compressive stress, C, that arises in the left pier:

$$C = \frac{1}{2}\left(\frac{F}{\sin\varphi} + \frac{W}{\cos\varphi}\right). \tag{TA10.1}$$

where W is the deck weight and φ is the bridge pier angle shown in figure 5.21a. For large earthquake force F and small pier angle φ (roughly 10° for the Lion's Gate Bridge), you can see that the compressive force can be large. The pier might buckle because of a horizontal jolt from an earthquake.

Now let us compare this result with the compressive force obtained when we loosen the pier connections to the ground. This is modeled in figure 5.21b; the pier is permitted to rock on its foundations. The loosened connectors allow the right pier to be raised off the ground; had the force been in the opposite direction, then the left pier is the one that would lift. What is the compressive force in the left pier now? I will skip the derivation;[2] the result is

$$C = F\sin(\varphi - \theta) + W\cos(\varphi - \theta), \tag{TA10.2}$$

where θ is the tilt angle of the rocking bridge. This tilt angle must not be greater than φ or the bridge will topple over (but recall that for the real bridge the nuts are loosened, not removed). Equations (TA10.1) and (TA10.2) are plotted in figure 5.22.

Note 11: Hydroelectric Dam Flow Rate

The energy of a mass, M, of water stored in a dam reservoir at height H is $E = MgH$. Recall that g is the acceleration due to gravity, so that Mg is the water weight. The dam's power output, P, is the rate of change of energy with time: this is the definition of power. So, from the equation for energy, we see that the output power is the rate of change of water mass, multiplied by a constant, gH.

1. Technically, this sleight of hand is simply a change of reference frames. What matters to us are the stresses that arise within the pier structure, and these do not depend upon our external reference frame. Thus, you can see that a force, F, to the right, applied to the two pins at the triangle base, will induce a compressive stress, C, in the left pier, as shown in fig. 5.21a. If instead we consider a force to the left applied at the apex, we get the same compressive force on the left pier.

2. You may calculate it yourself by noting that the only weight in the system is located at the apex, as shown, and by considering the components of the two forces along the direction of the left pier.

Now we know that mass is density multiplied by volume, $M = \rho V$, where ρ is water density, and so rate of change of water mass is just density times rate of change of volume. But rate of change of volume is the water flow rate, which I will denote f. Putting all this together we find that output power is given by

$$P = \rho g H f. \tag{TA11.1}$$

So if we know the output power of a dam and the height of its reservoir, then we can calculate flow rate from equation (TA11.1).

For example, water at Hoover Dam typically falls about 520 feet as it passes through the dam. That is to say, the water enters the hydroelectric power plant intake, on the upstream side of the dam, and then falls 520 feet before passing through the turbines and emerging on the downstream side of the dam. The potential energy of the water due to its height, or more accurately the power generated as it falls, is converted by the turbines into useful electrical power. So, for Hoover Dam $H = 520$ ft. The power output of this dam is about 2,000 megawatts; by substituting numbers for water density and the acceleration due to gravity, we obtain a flow rate, f, of 340,000 gallons per second. That's a lot of water. In fact we expect the true figure to be a little larger because no turbine is 100% efficient, and so the Hoover Dam turbines will not be able to convert all of the water power into electricity. In other words, the flow rate has to be a little larger than we calculated to produce the rated power output. In fact, the discharge spillways of Hoover Dam can deal with a maximum flow rate of about 500,000 gallons per second. The dam does not operate at peak discharge rate all the time, so you can see that our simple estimate of flow rate is pretty good.

Note 12: Dam Cross-Section Calculation

The horizontal ground reaction force X that acts upon the dam of figure 6.6a is just F, acting at pin1 in the direction shown. No surprise here; the ground is opposing the force exerted by the water on the dam, so the dam does not move. The vertical ground reaction forces, acting at both the foundation pins, add up to W. Again, no surprise: the ground is supporting the weight of the dam above it. It is interesting to look at the vertical ground reaction forces in a little more detail. We find that $Y_1 = \frac{3}{4} W - \frac{1}{7} F \cot \varphi$ and so $Y_2 = \frac{1}{4} W + \frac{1}{7} F \cot \varphi$. Now Y_2 acts at the downstream end of the foundation, and because it is positive, it shows that the bedrock is pushing upward, holding the dam in place. However, Y_1 is not necessarily positive. For small φ it can be negative—and if so, there is an uplift force tearing the upstream foundation from the bedrock. This is bad news because such an uplift will weaken the dam and might lead to water seeping underneath or the dam overturning. We want to avoid this possibility, and so we must choose an angle φ that is large enough to ensure that Y_1 is positive. For our truss model, we need $\varphi > 16°$. Real gravity dams will have a downstream slope angle exceeding this minimum, to include a safety factor.

Here is how this value for φ arises. The water pressure that acts upon the upstream face at a depth x beneath the water surface is $P = \rho g x$, where ρ is water density. From this equation we can show that the total force of water pushing against the dam is $F = ½ \rho g w H^2$, where w is the dam width (into the page, in fig. 6.6) and H is reservoir height, or total water depth. The dam weight is $W = ½ \rho_c g w H^2 \tan\varphi$, where ρ_c is the density of concrete. So, $F = W(\rho/\rho_c)\cot\varphi$. Substituting into our expression for Y_1 and requiring Y_1 to be positive leads to the constraint $\tan\varphi > 2\sqrt{\rho/21\rho_c}$. The density of concrete is about 2.4 times the density of water, and so $\varphi > 16°$.

Note 13: Pancaking

A simple calculation gives us the time it takes for a pancaking building to collapse. Let us ignore air resistance—the apartment is built in a vacuum—and assume that the strength of each floor is exactly the same. Let us say that each floor can resist a momentum of magnitude p, meaning that if the floor above slams into it with momentum less than p, then the lower floor will not collapse.[3] If a floor is hit with momentum exceeding p, then it collapses. In figure 7.3 I assume that the top floor gives way suddenly and falls freely under gravity; it is initially motionless but accelerates to a speed $v_1 = \sqrt{2gh}$, where h is the height of each story—the distance between floors. This top floor strikes the floor below, causing both to collapse with an initial speed $v_1 - p/m$, where m is the mass of the top floor. Both these floors then accelerate under gravity and fall onto the floor below with speed v_2, causing it to collapse with initial speed $v_2 - p/2m$. (The factor of 2 arises because two floors, with total mass $2m$, are falling.) The collapse continues like an avalanche, picking up speed as it goes. Let us say that the maximum momentum that each floor can resist is $p = mv_1$, so that the top floor only just causes the next-to-top floor (the $k = 1$ floor, in the notation of fig. 7.3) to collapse. The $k = 2$ floor will collapse more readily because it has twice the momentum striking it. The $k = 3$ floor offers even less resistance because it has the mass of three higher floors falling upon it, and so on. So, all the floors collapse. Had the top floor momentum, mv_1, been less than p, no progressive collapse would have occurred. Either no collapse occurs (apart from the top floor) or all the floors collapse.

3. Newton considered that dynamics were guided by the action of forces, and so, for the physicists among you, I should explain the collisions in terms of forces. Let us say that when a floor strikes the floor below, the collision takes place over a short time interval t, and so the maximum force that the lower floor can resist is p/t.

How long does it take for all the floors in our tower block to collapse? We have assigned the minimum possible top floor momentum that can initiate a progressive pancake collapse, and so we can anticipate that the collapse time is the *maximum* time possible (without air resistance). If the top floor slammed down on the $k = 1$ floor with more momentum than we have assigned, the collapse time would be less. I hope that this is clear to you; if not, stare at figure 7.3 for a while and you will see what I mean. Applying Newton's laws we find that the collision speeds, v_k, of figure 7.3 can be calculated iteratively:

$$v_{k=1} = \sqrt{\left(v_k - \frac{v_1}{k} \right)^2 + v_1^2} \, . \tag{TA13.1}$$

We know v_1 and so can calculate v_2 and then v_3, and so on. We can now calculate the total collapse time—the interval between the time the top floor begins its fall and the time the whole floor structure strikes the ground. We find it to be given by

$$\frac{T_K}{T_f} = \frac{1}{\sqrt{K}} \left[\frac{v_K}{v_1} + \sum_{k=1}^{K-1} \frac{1}{k} \right] . \tag{TA13.2}$$

Equation (TA13.2) is plotted in figure 7.4. In the equation, K is the total number of floors in the apartment block, T_K is the collapse time for the apartment block, and T_f is the free-fall collapse time. Free-fall time is simply the time it would take the top floor to fall if there were no other floors below it to impede its rush to the ground. (For K floors the free-fall time is $T_f = \sqrt{2Kh/g}$.) From figure 7.4 you can see that the collapse times are never very much greater than free-fall collapse time. In fact, it is easy to show that for an apartment tower with infinitely many floors, the ratio is $T_\infty / T_f = 1$.

Glossary

This glossary is not a comprehensive dictionary of structural engineering and architectural terms. I include only those terms that are used in the text and are not vernacular. Thus, for example, you will find *frog* but not *brick*.

Abutment. A structure that receives the sideways thrust of an arch or a vault.

Aisle. A part of the interior of a church next to the nave, often separated from it by columns.

Arcade. A series of side-by-side arches (like the loops in a lowercase **m**) supported by columns or piers.

Arch dam. A dam shaped like an arch that relies on material strength instead of weight to resist the horizontal force of water.

Area moment of inertia. A property of a structure's shape that can be used to estimate the structure's resistance to bending.

Axial force. A force that acts along the length of a structural element.

Barrel vault. A semicircular arched vault.

Beam. A structural member designed to resist bending.

Bending/flexural strength. A measure of the resistance to bending of a material of specified shape.

Bonding. The cementing together of bricks with mortar. There are numerous ways of combining bricks: English bond, Scottish bond, Flemish bond, herringbone, basket, etc.

Box girder bridge. A fixed bridge formed usually by welding four or more steel plates to form a box.

Bowstring arch bridge. An arch bridge with a tie that resists the horizontal thrust of the arch and forms a chord of the bridge.

Brittleness. A property describing a material that fractures easily when stretched or bent.

Buckling. The bending of a column that is subjected to a compressive load.

Buttress. A structure built against a wall to resist the horizontal thrust from a roof or an arch.

Cantilever. A truss or other structural member that projects beyond its supporting wall or column.

Cast iron. A brittle form of iron with a low melting point; it had many applications in the Industrial Revolution and is still widely used in cookware and other applications today.

Catenary. The curve of a hanging chain.

Centering. The supporting frame that holds up an arch during construction.

Chord. A row of members at the top or bottom of a truss bridge.

Clerestory. A high wall with a row of windows at the top, above a roofline.

Cofferdam. An enclosure to keep water out of an area of construction, such as a dam foundation or bridge pier.

Column. A vertical post carrying compressive forces.

Composite. A material formed of two or more components (still retaining their individual identity within the composite) that combines the strength of its constituents.

Compression. A squashing force.

Constant angle arch dam. A thin-walled dam with a shape that forms part of a cone.

Constant radius arch dam. A thin-walled dam with a shape that forms part of a cylinder.

Corbel. An architectural bracket projecting from a wall.

Cornice. An ornamental feature separating different floors of a building.

Corrosion. Destruction (particularly of metal) by chemical means.

Course. A layer of bricks.

Creep. The gradual deformation of a structure due to an applied force.

Critical column length. The length above which a column will fail by buckling rather than by compression.

Dam. A structure that is built to retain water, rather than to prevent water flow.

Damping. The reduction of oscillation amplitude due to dissipative forces such as friction.

Dead load. The weight of a structure.

Ductility. The ability of a material to stretch or bend without fracturing when subjected to a tensile stress.

Dynamic load. A time-dependent external force that leads to elastic deformation of a structure.

Elastic deformation. Reversible deformation of a material.

Embankment dam. A bank of earth or stone built to retain water.

Extrados. The exterior curve or surface of an arch.

Fan vault. A vault with ribs that radiate from a common point, often a column.

Fatigue. Progressive, localized structural damage to a material that is subjected to cyclic loading.

Five-minute rule. A rule of thumb for judging structural design: if the structure remains intact for 5 minutes, then the design is good.

Flexural stiffness. A measure of the resistance of a structure to bending.

Forced oscillation. The oscillations of a body that is subjected to an external periodic driving force.

Frog. The indentation on the broad face of a brick.

Funicular. A structural shape adopted so that the only load that acts upon it is compression (or tension).

Girder. A beam, nowadays usually made of steel and with an I-shaped or rectangular cross-section, that forms the main horizontal support of a structure.

Gothic arch. An arch with a pointed apex.

Gravity dam. A concrete dam that resists the pressure of water by virtue of its weight.

Groin vault. A vault formed by the intersection of two (or more) barrel vaults.

Hammer beam roof. An open timber roof formed of trusses with no tie beams, creating extra space.

Hardness. A material property describing resistance to permanent deformation.

Haunch. The side of an arch between keystone and pier.

Header. The short end of a brick.

Herringbone. A brickwork pattern consisting of rows of short parallel lines, with direction alternating row by row.

Horseshoe (Moorish) arch. An arch with a curve that extends beyond a semicircle.

Hypar (hyperbolic paraboloid). A saddle-shaped curved surface formed from straight members, used architecturally to form a roof.

Hyperboloid. A radially symmetric curved surface formed from straight members, used architecturally to form towers.

Implosion. Destruction via explosives, in such a way that the structure falls inwards and onto its footprint.

Impost. The lowest voussoirs of an arch; the stones upon which the arch rests.

Influence line. The sequence of forces that act upon the member of a truss structure that is subjected to a moving load.

Intrados. The interior curve or surface of an arch.

Keystone. The central, uppermost stone of a thin arch: the highest voussoir.

King post roof. A timber roof structure in which a single vertical supporting post connects a tie beam with rafters.

Lintel. A horizontal structural member that spans an opening, as over a door or window.

Live load. An external force that acts upon a structure—for example, the weight of people on a bridge.

Magnetorheological (MR) fluid. A "smart" fluid which increases its viscosity when subjected to a magnetic field.

Malleability. The ability of a material to distort without fracturing when subjected to a compressive stress.

Member. A rigid structural element.

Modulus. A parameter describing a mechanical property of a material. **Young's modulus:** parameter describing tensile elasticity. **Shear modulus:** parameter describing rigidity.

Mortar. A mixture of cement, water, and fine aggregate used to bind structural blocks, such as bricks, together.

Natural frequency. The frequency at which an object will vibrate freely after an initial push.

Nave. The central hall or passageway of a church.

Neutral layer. A layer of a structural member subjected to bending that is neither stretched nor compressed.

Orthotropic. Referring to a material whose properties are independent of direction—for example, one whose tensile strength does not depend upon the direction of the applied force.

Pancaking. The rapid collapse of a tall building from the top; falling debris accelerates the process.

Parabola. The shape of the curve of suspension bridge cables.

Pendentive. A triangular section of vaulting lying between the rim of a dome and the supporting arches.

Pin. A hinge connecting truss members.

Plastic. A material property describing a material's ability to deform without breaking.

Pony truss bridge. A bridge consisting of two parallel trusses on either side of the deck, with no covering on top. The trusses can therefore be low, typically waist-high.

Pozzolana. Roman mortar.

Prestressed concrete. A concrete beam or other structural element with a compression stress induced in it before use. Prestressed concrete beams can support much greater loads than a plain concrete beam of the same size.

Purlin. A roof truss member lying perpendicular to the rafter, running along the length of the roof.

Queen post roof. A timber roof structure in which two vertical supporting posts connect a tie beam with rafters.

RDX. A high explosive, cyclotrimethylenetrinitramine, used in controlled demolitions.

Rebar. Reinforcing steel bar.

Reinforced concrete. Concrete strengthened with rebar.

Resonance. The increase in the oscillation amplitude of a mechanical or other physical system that is subjected to a periodic force with a frequency close to the system's natural frequency.

Ribbed vault. A vault in which the surface is strengthened by a framework (sometimes diagonal) of arched ribs.

Rule of the middle third. A rule of thumb for choosing the thickness of a supporting column or wall: the vector representing the sum of forces that act on a column or a wall should pass through the middle third of the base.

Segmental arch. An arch with an arc that is smaller than a semicircle.

Setback. A shelf or a recess in a wall.

Shear force. Force causing a material to skew; a plane in the material displaces parallel to itself.

Shear modulus. The ratio of shear stress to shear strain.

Shear wall. A wall consisting of braced panels that resist horizontal shear; used to provide stiffness in buildings.

Soffit. The underside of an architectural structure such as an arch, a beam, or a ceiling.

Spandrel. The (often filled) space between an arch and the structure it supports.

Springer. The first voussoir above the impost of an arch; the second stone from the bottom.

Statically determinate. Referring to a structure for which the static equilibrium equations are sufficient to determine all the internal and reaction forces.

Statically indeterminate. Referring to a structure for which the static equilibrium equations are not sufficient to determine all the internal and reaction forces. Additional structural properties must be invoked for this.

Steel alloy. A homogeneous mixture of iron and carbon (steel) plus another metal.

Stiffness. The resistance of a material to deformation.

Strain. The deformation of a material that is subjected to stress.

Stress. The force per unit area applied to a material.

Stretcher. The long side of a brick.

Surcharge. An additional load, such as that atop an arch.

Tensegrity. Tension integrity, a term used to refer to a structure that maintains its form because of a balance between members in tension.

Tensile, tension. A pulling or stretching force.

Tie beam. The horizontal lower chord of a truss roof.

Torsion/torque. A twisting force.

Toughness. A property describing a material's ability to absorb energy without being destroyed.

Trabeated. A term descriptive of a building constructed using beams or lintels, not arches or vaults.

Transept. The transverse portion of a building that cuts its main axis at right angles; in a cruciform church, the "arms" of the cross shape.

Trestle. A framework of straight members supporting a bridge.

Truss. A stable assemblage of straight and rigid structural members joined at their ends.

Ultimate strength. The maximum stress to which a material can be subjected before failing.

Vault. An arch that is deeper than it is wide; an arch tunnel.

von Karman effect. Oscillations induced in a bluff body by fluid flowing across it.

Voussoir. One of the wedge-shaped stone blocks that define the curve of a masonry arch.

Web. The internal members of a truss structure.

Yield strength. The maximum stress to which a material can be subjected before plastic deformation occurs.

Zero-force member. A truss member that is subjected to neither compression nor tension.

Bibliography

This bibliography lists primary and secondary specialized sources covering material specific to each chapter under separate chapter headings. A number of more general references that have been useful in more than one chapter are listed at the beginning.

General References

Billington, David P. *The Tower and the Bridge: The New Art of Structural Engineering.* Princeton: Princeton University Press, 1985.

———. *The Art of Structural Design: A Swiss Legacy*. New Haven: Yale University Press, 2003.

These Billington titles are two of many books by an authority who emphasizes art in structural engineering. Describes the work of Telford, Brunel, Eiffel, Roebling, Freyssinet, Ammann, and especially Maillart.

Chrimes, Mike. *Civil Engineering, 1839–1889*. Stroud, UK: Sutton, 1991.

A nontechnical account of the achievements of civil engineering in Victorian Britain.

Encyclopaedia Britannica. CD 98 Standard Edition. 1998. S.vv. "Arch," "Bridge," "Dam," "Dome," "Vault."

James, Peter, and Nick Thorpe. *Ancient Inventions*. New York: Ballantine, 1994.

Macaulay, David. *Building Big* Website. www.pbs.org/wgbh/buildingbig.

Macaulay's PBS TV series website covers the same ground as this book, at a less technical (junior high school) level.

Mason, Stephen F. *A History of the Sciences*. New York: Collier, 1962. Pp. 503–12.

A comprehensive history, with a section on key engineering developments that places them in historical context.

Microsoft Encarta Encyclopedia. CD ROM. 2004. S.vv. "Arch and Vault," "Bridge," "Dam," "Dome."

Petroski, Henry. *The Evolution of Useful Things*. New York: Random House, 1992.

Includes an interesting discussion on engineering failure.

Reid, Esmond. *Understanding Buildings: A Multidisciplinary Approach*. Cambridge: MIT Press, 1988.

Salvadori, Mario. *Why Buildings Stand Up*. New York: W. W. Norton, 2002.

Symon, Keith R. *Mechanics*. Reading, MA: Addison-Wesley, 1960.

An undergraduate physics text with an unusually clear account of simple beam theory.

CHAPTER 1. Building Blocks

Cotterell, Brian, and Johan Kamminga. *Mechanics of Pre-industrial Technology*. Cambridge: Cambridge University Press, 1990.

Gordon, J. E. *The New Science of Strong Materials*. London: Penguin, 1976.

Millais, Malcolm. *Building Structures*. Oxford, UK: E. and F. N. Spon, 1997.

Morgan, W. *The Elements of Structure*. London: Pitman, 1964.

Neuringer, Joseph L., and Isaac Elishakoff. "Interesting Instructional Problems in Column Buckling for the Strength of Materials and Mechanics of Solids Courses." *International Journal of Engineering Education* 14 (1998): 204–16.

Raymond, Robert. *Out of the Fiery Furnace*. University Park: Pennsylvania State University Press, 1986.

Usher, Abbott Payson. *A History of Mechanical Inventions*. New York: Dover, 1982. Chap. 6.

CHAPTER 2. Truss in All Things High

Griggs, Frank. "Evolution of the Continuous Truss Bridge." *Journal of Bridge Engineering* 12 (2007): 105–19.

Johns Hopkins University Spaghetti Bridge Building Contest. Video. www.researchchannel.org/prog/displayevent.aspx?rID=4070&fID=569.

Matsuo Bridge Co. "Truss." www.matsuo-bridge.co.jp/english/bridges/basics/truss.shtm. 1999.

On its website the Matsuo Bridge Company provides simple and nontechnical explanations of various truss (and other) bridge designs. Links to discussions of other bridge designs appear at www.matsuo-bridge.co.jp/english.

Rower, Ken, ed. *Historic American Roof Trusses*. Becket, MA: Timber Framers Guild, 2006.

Sheppard, Charles, *Railway Stations*. New York: Todtri, 1996.

CHAPTER 3. Towers of Strength

In addition to the books and articles listed below, there is a large Internet resource on tall buildings. For example, most of the famous tall buildings and towers have their own websites.

Ali, Mir M., and Kyoung Sun Moon. "Structural Developments in Tall Buildings: Current Trends and Future Prospects." *Architectural Science Review* 50, no. 3 (2007): 205–23.

Chicago Tribune. "How Tall Is Too Tall?" (editorial), July 27, 2005.

Dupré, Judith, and Phillip Johnson. *Skyscrapers*. New York: Black Dog and Leventhal, 1996.

Fortmeyer, Russell. "The Skyscraper: Still Soaring." *Business Week*, August 17, 2007.

Nordenson, Guy, and Terence Riley. *Tall Buildings*. New York: Museum of Modern Art, 2003. Associated MOMA website: www.moma.org/exhibitions/2004/tallbuildings/index_f.html.

CHAPTER 4. Arches and Domes

Croci, Giorgio. "Seismic Behavior of Masonry Domes and Vaults: Hagia Sophia in Istanbul and St. Francis in Assisi." Keynote address, First European Conference on Earthquake Engineering and Seismology, Geneva, Switzerland, September 3–8, 2006.

Levy, Matthys. "The Arch: Born in the Sewer, Raised to the Heavens." *Nexus Network Journal* 8, no. 2 (2006): 7–12.

CHAPTER 5. A Bridge Too Far

Bonanos, Christopher. "The Father of Modern Bridges." *American Heritage of Invention and Technology* 8, no. 1 (1992): 8–20.

Brantacan Co. "Bridges." www.brantacan.co.uk/bridges.htm.

A detailed website with many links, discussing all aspects of bridges in an accessible style.

Buonopane, Stephen G. "The Roeblings and the Stayed Suspension Bridge: Its Development and Propagation in 19th-Century United States." In *Proceedings of the Second International Congress on Construction History*, ed. Malcolm Dunkeld et al., 1:441–60. Exeter, UK: Short Run Press, 2006.

Cassady, Stephen. *Spanning the Gate*. Mill Valley, CA: Squarebooks, 1986.

Clericetti, C. "The Theory of Modern American Suspension Bridges." *Minutes of Proceedings of the Institution of Civil Engineers* 60 (1880): 338–59.

Drewry, Charles Stewart. *A Memoir of Suspension Bridges*. London: Longman, 1832.

An account of the Broughton Bridge collapse, caused by marching soldiers, is given on pp. 92–94.

Hayden, Martin. *The Book of Bridges*. New York: Galahad, 1976.

McCullough, David. *The Great Bridge*. New York: Simon and Schuster, 1972.

Moran, Barbara. "A Bridge That Didn't Collapse." *American Heritage of Invention and Technology* 15, no. 2 (1999): 10–18.

NOVA (PBS). "Arch Bridge." *Super Bridge* Website. www.pbs.org/wgbh/nova/bridge/meetarch.html. October 2000.

An educational but not too technical account of bridge engineering, in this case examining the strength of arch bridges.

Troyano, Leonardo Fernandez. *Bridge Engineering: A Global Perspective*. London: Thomas Telford, 2003.

CHAPTER 6. Dam It

Duggal, K. N., and J. P. Soni. *Elements of Water Resource Engineering*. New Delhi, India: New Age, 1996. Chap. 15.

Jackson, Donald C. *Great American Bridges and Dams*. Hoboken, NJ: John Wiley, 1988.

The Story of Hoover Dam. http://video.google.com/videoplay?docid=-2361644829738750659.

This website shows an old Bureau of Reclamation movie about the construction of Hoover Dam. This half-hour film presents a rosy view of the project (no mention of lives lost or poor working conditions, or of the naming controversy, for example) but provides a good overview of the engineering achievements and of the benefits provided by the dam.

CHAPTER 7. The Bigger They Are, the Harder They Fall

Bažant, Zdenìk P., Jia-Liang Le, Frank R. Greening, and David B. Benson. "What Did and Did Not Cause Collapse of WTC Twin Towers in New York." Northwestern University Structural Engineering Report No. 07-05/C605c, available online at www.civil.northwestern.edu/people/bazant/PDFs/Papers.

Blanchard, Brent. "A Critical Analysis of the Collapse of WTC Towers 1, 2, and 7 from an Explosives and Conventional Demolition Industry Viewpoint." August 8, 2006. *ImplosionWorld.com* Website. http://www.implosionworld.com/news.htm.

Eager, Thomas W., and Christopher Musso. (2001) "Why Did the World Trade Center Collapse? Science, Engineering, and Speculation." *JOM: Journal of the Minerals, Metals, and Materials Society* 53, no. 12 (2001): 8–11.

Ferguson, Eugene S. "How Engineers Lose Touch." *American Heritage of Invention and Technology* 8, no. 3 (1993): 16–24.

ImplosionWorld. *ImplosionWorld.com* Website. www.implosionworld.com.

Many videos and still images of imploding buildings and other structures can be seen on this website.

Nantais, Chad. "Massive Building Implosion—Detroit." *YouTube* Website. www .youtube.com/watch?v=khD2gZkkSu0.

This YouTube video shows the controlled demolition on October 24, 1998, of the Hudson Building in Detroit, the tallest building so far brought down by this method.

Petroski, Henry. *To Engineer Is Human.* New York: Random House, 1992.

Sofge, Eric. "Delicate Art of Building Demolition." *Popular Mechanics*, November 2005.

Varieschi, Gabriele, and Kaoru Kamiya. "Toy Models for the Falling Chimney." *American Journal of Physics* 71, no. 10 (2003): 1025–31.

See also my comment on this paper in *American Journal of Physics* 74, no. 1 (2005): 82–83. Varieschi's falling chimney website is at http:// myweb.lmu.edu/gvarieschi/chimney/chimney.html.

Index